```
TJ          Stirling and
262           Vuilleumeir heat
.S75          pumps.
1990
```

$48.95

BAKER & TAYLOR BOOKS

REF

BUSINESS/SCIENCE/TECHNOLOGY DIVISION

The Chicago Public Library

TJ
262
.S75
1990

Received

Stirling and Vuilleumier Heat Pumps

Stirling and Vuilleumier Heat Pumps

Design and Applications

Jaroslav Wurm
John A. Kinast
Thomas R. Roose
William R. Staats

McGraw-Hill, Inc.

New York   St. Louis   San Francisco   Auckland   Bogotá
Caracas   Hamburg   Lisbon   London   Madrid
Mexico   Milan   Montreal   New Delhi   Paris
San Juan   São Paulo   Singapore
Sydney   Tokyo   Toronto

Library of Congress Cataloging-in-Publication Data

Stirling and Vuilleumier heat pumps: design and applications / Jaroslav Wurm . . . [et al.].
  p. cm.
Includes bibliographical references and index.
1. Heat pumps. I. Wurm, Jaroslav.
TJ262.S75 1990
621.402′5—dc20   90-6540
ISBN 0-07-053567-1   CIP

Copyright © 1991 by McGraw-Hill, Inc. All rights reserved. Printed in the United States of America. Except as permitted under the United States Copyright Act of 1976, no part of this publication may be reproduced or distributed in any form or by any means, or stored in a data base or retrieval system, without the prior written permission of the publisher.

1 2 3 4 5 6 7 8 9 0 DOC/DOC 9 8 7 6 5 4 3 2 1 0

ISBN 0-07-053567-1

The sponsoring editor for this book was Robert W. Hauserman, the editing supervisor was Jim Halston, the designer was Naomi Auerbach, and the production supervisor was Thomas G. Kowalczyk. It was set in Century Schoolbook by McGraw-Hill's Professional Publishing composition unit.

Printed and bound by R. R. Donnelley and Sons Company.

Information contained in this work has been obtained by McGraw-Hill, Inc. from sources believed to be reliable. However, neither McGraw-Hill nor its authors guarantees the accuracy or completeness of any information published herein and neither McGraw-Hill nor its authors shall be responsible for any errors, omissions, or damages arising out of use of this information. This work is published with the understanding that McGraw-Hill and its authors are supplying information but are not attempting to render engineering or other professional services. If such services are required, the assistance of an appropriate professional should be sought.

For more information about other McGraw-Hill materials, call 1-800-2-MCGRAW in the United States. In other countries, call your nearest McGraw-Hill office.

Dedicated to my parents, who gave me their unselfish guidance through their difficult lives.

JAROSLAV WURM

Dedicated with love and appreciation to my wife, Barbara, who supported me and gave me the time to develop these numbers, words, and pictures, and to God who made everything we explore.

JOHN A. KINAST

Dedicated to my loving wife, Rosmarie, and my caring children, Sarah, Robert, Peter, and Laura, who give purpose to all I do.

THOMAS R. ROOSE

Dedicated with love and appreciation to my wife, Ann, and to my children, Meredith and Andrew.

WILLIAM R. STAATS

Contents

Preface   xi

Chapter 1. Introduction   1

- Engine-Driven Heat Pumps   2
- Reference Cycles   4
- Cycles Based on Two-Phase Working Fluids   4
- Cycles Based on Single-Phase Working Fluids   5
- References   7

Chapter 2. Heat Pump Theory   9

- General Heat Pump Operation   9
- General Categories of Reference Cycles   13
- Performance Parameters   14
  - Coefficient of Performance   15
  - Refrigeration Efficiency   16
  - Specific Volumetric Heating and Cooling Capacity   17

Chapter 3. Reference-Cycle Thermodynamics   19

- Ideal Reference Cycles   20
  - Carnot Cycle   20
  - Stirling and Ericsson Cycles   22
- Other Theoretical Reference Cycles   25
  - Lorenz Theoretical Cycle   26
  - Clausius-Rankine Theoretical Cycle   27
  - Brayton Theoretical Cycle   28
- Combinations of Cycles   29
  - Staged Refrigerators   30
  - Combined Engine-Refrigerator Cycles   30
  - Vuilleumier Cycle   31
- Summary   33
- References   34

Chapter 4. Overview of Integrated Heat Pumps Concepts   35

- Desirable Attributes of Heat Pumps   35
- Classification of Heat-Activated Heat Pumps   36

viii   Contents

Advantages of Integrated Heat Pumps	38
Selection of Concepts for Analysis	39
Origin and Development of Integrated Heat Pump Concepts	41
Traditional Vuilleumier Heat Pump	41
Vuilleumier Heat Pump With Internal Heat Exchangers	43
Duplex Stirling Heat Pump	43
Balanced-Compounded Stirling Heat Pump	44
Balanced-Compounded Vuilleumier Heat Pump	44
Ericsson-Ericsson Heat Pump	44
References	45

Chapter 5. Description of Stirling-Stirling and Vuilleumier Heat Pumps   47

Traditional Vuilleumier Heat Pump	48
Description	48
Operation	49
Vuilleumier Heat Pump With Internal Heat Exchangers	51
Description	51
Operation	52
Duplex Stirling Heat Pump	52
Description	52
Engine Operation	54
Refrigerator Operation	56
Gas Spring Operation	57
Balanced-Compounded Stirling Heat Pump	58
Description	58
General Operation	58
Engine Operation	59
Refrigerator Operation	61
Balanced-Compounded Vuilleumier Heat Pump	63
Description	63
Engine Operation	65
Refrigerator Operation	67
Ericsson-Ericsson Heat Pump	68
Description	68
Operation	69
Gas Spring Operation	71
Summary	72
References	74

Chapter 6. Analytical Methodology   75

Stirling-Cycle Analysis	75
Recent Analytical Developments	80
Techniques Used in This Analysis	82
Definition of Operating Conditions	83
References	85

Chapter 7. Comparisons of Stirling-Stirling and Vuilleumier Heat Pumps   87

General Analysis Method	87
Assumptions	88
Analytical Simulation	88
Common Operating Conditions	89

Ideal-Cycle Analysis	90
Thermal-Compression Vuilleumier Concepts	90
Stirling-Stirling Concepts	92
Mechanical-Compression Vuilleumier Concept	97
Ericsson-Ericsson Concept	99
Effect of More Realistic Design Factors	101
Dead Space	101
Regenerator Inefficiencies	104
Varying Heat Flux for Isothermal Operation	113
Mechanical Energy Storage	116
Summary	120
Thermal-Compression Concepts	120
Mechanical-Compression Concepts	121
Free-Piston Configuration Considerations	123
References	123

Chapter 8. Analysis of Real Integrated Heat Pumps — 125

Recent Hardware Development Programs	125
Philips Research Laboratories	125
Professor Eder's Vuilleumier Heat Pumps	126
Sunpower's Duplex Stirling Heat Pumps	129
University of Dortmund Vuilleumier Heat Pumps	132
Sanyo Vuilleumier Heat Pump	134
Design Optimization	136
Analysis	136
Effects of Realistic Heat-Exchanger Design	140
Design Optimization Tools	141
Model Validation	147
Conclusions	147
References	148

APPENDIX A Thermal-Compression Vuilleumier Heat Pump Program — 151

APPENDIX B Stirling-Stirling (Piston-Piston) Heat Pump Programs — 161

APPENDIX C Stirling-Stirling (Displacer-Piston) Heat Pump Programs — 179

APPENDIX D Mechanical Compression Vuilleumier Heat Pump Program — 199

APPENDIX E Ericsson-Ericsson Heat Pump Program — 211

APPENDIX F Thermal-Compression Vuilleumier Heat Exchanger Analysis Program — 223

APPENDIX G Glossary — 241

Index — 249

Preface

There is widespread concern that the use of fossil fuels has been too profligate. Fears of fossil fuel depletion pervaded much of the 1970s and 1980s. Now we are beginning to learn that our use of fossil fuels is producing carbon dioxide faster than the atmosphere and oceans can accommodate it.

With no complete substitute at hand for fossil fuels, it is increasingly important to use them as effectively as possible. This book approaches one aspect of that use: heating and cooling of occupied buildings. The use of energy for space conditioning or comfort heating and cooling is now considered a necessity in most industrialized countries. In the United States, space conditioning accounts for about one-third of all the fossil fuel consumed for purposes other than transportation. Fuel use for space conditioning could be reduced substantially if heat pumps, with their substantial efficiency advantages, were more widely used.

The term heat pump is used throughout this book to describe the combination of the driving unit, or engine, and the energy consuming device, or refrigerator, which absorbs heat at a lower temperature and discharges it at a higher temperature. Heat pumps are used for both heating and cooling. Strictly speaking, the electrically driven heat pumps used to heat and cool buildings only fit the above definition of complete heat pumps when the large engine at the electric power station is considered a part of them.

This book deals with six examples of engine-driven heat pumps. Both the engine segment and the refrigerator segment of these heat pumps are based on Stirling cycles or similar thermodynamic cycles. Stirling-cycle external combustion engines, in principle, offer higher efficiencies and better maintenance characteristics than internal combustion engines. Stirling refrigerators also have potential performance benefits. These heat pump combinations are not commercially available; and, for some of them, prototype equipment has not even been built. They all need further development, particularly to reduce their cost. However, they all have one important characteristic that can lead to cost reduction. They combine the engine and refrigerator segments and share components between them. This intermingling of the engine and refrigerator functions leads to the term *integrated heat pump,* which is used throughout the book.

Many combinations of different engines and refrigerators are possible, and the ingenuity of several inventors has resulted in the development of many different prototypes. None of these prototypes have entered commerical production. The perceived advantages of the new designs have not yet been impressive enough to overcome the large capital investments required for mass production. To convert a promising idea into effective prototype hardware is costly, both in R&D funds and time. The R&D would cost less if we could calculate the potential performance of the prototype equipment and critically compare it to other configurations before building it. Such an analysis could also suggest better configurations and help in prototype design by showing how certain components limit system performance. This would allow the designers to balance performance improvement against the cost of better components.

Unfortunately, integrated heat pumps have not received the same comprehensive literature coverage as similar concepts with separate cycles that are linked mechanically or electrically. The technical literature does not contain much information on integrated heat pumps, and most engineers are not familiar with their analysis. Furthermore, the few engineers who are experts in this field have evolved detailed analyses of the performance of the specific equipment concept they are developing. Although these mathematical models are precise, they do not allow even-handed comparison with other concepts. This makes it difficult to choose objectively among the various concepts that have been proposed.

We intend to help remedy that situation by documenting consistent methods for analyzing the performance of this equipment. We focus on six concepts that integrate the engine and refrigeration cycles, all based on Stirling, Ericsson, and Vuilleumier cycles. The six heat pump concepts are examples of how to apply the analysis methods. In addition to comparing the performance of these concepts, the analytical methods give inventors and designers a way to quantitatively predict the performance capabilities of new systems and their components. Such evaluation can also lead to more confident R&D decisions through better assessment of the potential success of specific configurations. For the configurations that warrant substantial R&D, the techniques will improve the design of prototype equipment. Ultimately, we hope our readers will use the methods presented to devise even better integrated heat pump concepts.

A brief survey of engine-driven heat pump technologies in Chap. 1

delineates the scope of this book. It explains the usefulness of engine-driven heat pumps in general and integrated heat pumps in particular. Chapter 2 describes the thermodynamic basis for analyzing heat pump performance, and Chap. 3 extends that basis and applies it to heat pump concepts that integrate the engine and the refrigerator. Chapter 4 explains the factors that are important for commercial success of a heat pump concept and how these factors affect the selection of concepts for development. It also illustrates the advantages of the six specific embodiments of integrated heat pumps, which will be analyzed in subsequent chapters. Chapter 5 describes these six embodiments more fully and explains why we selected them for analysis. Chapter 6 outlines many of the methods that others have used to analyze Stirling cycle equipment performance and the methods used in this book to analyze integrated heat pump performance. Chapter 7 gives the results of our comparative analysis of the six embodiments. Chapter 8 adds a description of equipment that is now under development and outlines the analysis of components, such as heat exchangers, which are particularly important to the performance of integrated heat pumps.

Appendices A through E contain documentation and listings of the computer programs used in the analysis whose results are reported in Chap. 7. These programs define exactly how the analysis was performed, and they will be a guide to others who want to analyze similar integrated heat pump concepts.

Appendix F documents an example of a computer program that extends the above analyses to predict how realistic heat exchanger design affects heat pump performance, as described in Chap. 8. Mr. Marek Czachorski developed the program shown in Appendix F and the performance calculations shown in Chap. 8. We appreciate his consent to use them here.

This book stems from research performed at the Institute of Gas Technology, funded by the Gas Research Institute. We are grateful for the cooperation of these companies in allowing us to include the results of this research. We also appreciate the help of those researchers and companies who have produced research and prototype machines of practical value. Several of them gave us permission to describe their equipment and activities in this book. Without these contributions from Philips Research Laboratories, Sunpower, Inc., the University of Dortmund, the University of Munich, Professor Franz X. Eder, and Tokyo Gas Company, this book would have only academic interest.

J. Wurm
J. A. Kinast
T. R. Roose
W. R. Staats

Stirling and Vuilleumier Heat Pumps

Chapter 1

Introduction

Ventilation and ice storage were the only available ways to cool buildings until the mid-nineteenth century. The invention of a variety of engines which could drive refrigeration systems in the nineteenth century made space cooling technically possible, but practical prime movers (engines and electric motors) were not available until the twentieth century. By then, electric power generation and distribution systems had developed to the point where they could dominate space-cooling technology. Even now, the lower costs and maintenance requirements and greater convenience of electric motors usually outweigh the higher cost of electric energy.

For most of the world's population, comfort space conditioning is still limited to fresh air ventilation and heating of the living and working environment by combustion equipment. Yet, world opinion reflects increasing concern over the cost—both economic and environmental—of fossil fuels. There is an increasing desire to meet space-heating needs by more efficient fuel use. In principle, heat pumps offer the prospect of efficiencies that cannot be matched by any other means. The technology developed for comfort cooling has been adapted for comfort heating of buildings. Like cooling equipment, these heat pumps also use electric power.

Although electric power dominates this technology, the potential advantages of alternative power sources still intrigue the research community, and research and development (R&D) is continuing on small engines and other approaches. Much of this R&D has been well covered in the literature, but one important segment—integrated heat pumps based on Stirling, Ericsson, and Vuilleumier cycles—has not. This book provides a consistent description of this family of alternative cooling and heating concepts and methods for analyzing their performance. We hope that the availability of this information will encourage others to consider this potentially important class of equipment when developing new concepts for practical heating and cooling systems.

Engine-Driven Heat Pumps

As early as 1853, Sir William Thomson, Lord Kelvin, derived and documented the fact that engine-driven heat pumps could be designed.[1] In articles published in the *Cambridge and Dublin Mathematical Journal*, this great thermodynamicist described the use of an engine-driven heat pump and calculated its efficiency advantage over simple combustion. The mechanics of his concept may not have been practical, but his prediction of its potential usefulness was prescient. Although the heat pumping principle has been understood for well over a century, most buildings are still heated by combustion, with an efficiency well below 0.9. Commercially available vapor-compression heat pumps driven by electric motors offer better heating efficiencies for a range of conditions. They have disadvantages and limitations, but they are still an important step toward reaching the space-conditioning performance potential envisioned by Lord Kelvin.

Small engines at the building being heated can be an attractive alternative to electric power, which, in essence, uses a large engine at a central electric power station. The exhaust heat from a local small engine can augment the heat pump output during the heating season. In climates with significant heating loads, the annual fuel use for heating and cooling combined can be lower with heat pumps powered by small engines than with those powered by electricity. To be practical, however, this requires small engines which are efficient enough, are inexpensive, and have reasonable maintenance requirements. The engines must also have a far longer life than is normally required in vehicles. Most commercially available internal combustion engines meet these requirements only in sizes well above those needed for residences and small commercial buildings. However, those buildings are important because, in total, they consume much of the energy used for space heating.

As early as the 1930s, attempts were made to develop small-engine-driven air-conditioning and heat pump equipment. The Coleman Company field-tested such equipment in 1956, but it never reached the market. However, the recent remarkable diversity in new prime mover development has led to the development of many new engine-driven heat pumps covering a range of sizes. This worldwide activity has resulted in some successful marketing of residential and small commercial-size units in Japan and larger commercial-size units in Germany.

External combustion engines hold the promise of circumventing the disadvantages of small-engine-driven heat pumps. External combustion engines can have inherently higher efficiency than internal combustion engines, and their heat-rejection characteristics are amenable to recovering the exhaust heat for space heating. External combustion engines can have longer life and need less maintenance because the products of combustion do not contaminate the engine's internal working surfaces. The invention and initial reduction to practice of external combustion engines, originally called hot-air engines, predates that of their internal combustion counterparts, but eventually internal combustion engines supplanted them. The supporting technology needed for efficient, competitive external combustion engines was not available at the turn of the twentieth century, when engine development for automobiles was in its formative stage. Achieving high efficiency and adequate power from external combustion engines requires heat-exchanger designs and construction materials that only recently became readily available. In particular, a lack of materials that allow high-temperature heat inputs has been a critical limitation.

A heat pump needs a refrigerator as well as an engine. Ordinary open-shaft vapor-compression equipment would seem to be the best choice because it is already developed and has been accepted in commercial applications. However, coupling this refrigerator with an engine leads to high overall equipment costs. In most applications, the lower operating costs of such a system do not compensate for the higher equipment cost. This is a compelling reason to look into other engine options (such as the external combustion engines mentioned above), other refrigerator concepts, and, ultimately, novel combinations of engines and refrigerators.

Open-shaft vapor-compression refrigerators are not the only refrigerators that can be used in engine-driven heat pumps. For example, refrigerators which use a working fluid that does not change phase, such as air, offer advantages in cost, performance, and environmental acceptability. Stirling and Vuilleumier heat pumps are in this category; but, despite their potential advantages, there is only limited commercial experience with their use.

Some refrigerators whose working fluids do not change phase could be attractive for comfort heating because of their inherently high potential efficiency over a wide range of outdoor temperatures. Their disadvantage is that achieving such efficiency requires high-performance hardware with many components. The high cost of this complex hardware has prevented widespread application of this type of heat pump, so cost reduction is very important. A first step in reducing costs would be to simplify the hardware by combining component functions, and the most obvious candidates for combination are the heat exchangers. Another simplification is to integrate the engine function with the heat pump function, eliminating power-transmitting equipment. One simpler combination is known as the Vuilleumier heat pump, named after its inventor, who received U.S. patents in 1917 and 1918.[2]

Reference Cycles

The above discussion described engines and refrigerators in broad, general terms. However, the following chapters require a fully developed thermodynamic frame of reference for comparative analysis of integrated heat pumps. This section covers the same concepts as the previous section, but uses more formal categories and specific, precise definitions based on thermodynamic reference cycles. We begin to develop this frame of reference here and continue the development in Chaps. 2 and 3, defining all the cycles of potential interest and eliminating from consideration those that have either proven impractical or are well described in the literature. However, this book does not teach basic thermodynamics. The reader should already understand the basic thermodynamic cycles which, at least in theory, describe both power-producing equipment (engines) and power-consuming equipment (refrigerators). Readers who are not familiar with the common power cycles should consult general thermodynamics texts.[3,4]

Two categories of cycles are described below: those in which the working fluid changes phase from gas to liquid, and those whose working fluid remains in the gas phase. The former are commonly known as *vapor-compression cycles*. The cycles referred to apply to both engines and refrigerators.

Cycles Based on Two-Phase Working Fluids

When working fluids change phase during a cycle, they offer a closer approach to isothermal operation and they carry more energy per unit of fluid circulated than single-phase fluids. This is why equipment using vapor compression, or two-phase working fluids, dominates refrigeration technology. This equipment follows the Clausius-Rankine and Lorenz cycles. We will not analyze vapor-compression equipment because methods for analyzing its performance are freely available in the literature.

Clausius-Rankine machines which combine the engine cycle with the refrigeration cycle, using steam as the working fluid, have been proposed. They have been used in industrial refrigeration or process (noncomfort) air conditioning, and a few have reached commercial application. They are usually steam-turbine-driven vapor-compression machines (piston, screw, or turbine). Those steam-turbine machines that could be used for comfort air conditioning are still under development, and literature references to them are scarce. The most interesting ones use a single fluid in a double-loop configuration. The first designs were based on reciprocating piston expanders and compressors, but recent, more successful R&D has focused on turbomachinery.[5]

Another cycle that describes condensing working fluids, the Lorenz cycle, is becoming more important. Although it is not as well known as the Clausius-Rankine cycle, it actually gives a better description of the operation of vapor-compression equipment. It also allows better representation of cycles that use mixtures of working fluids. This cycle is also fully described in the literature.[6,7]

The Clausius-Rankine and Lorenz cycles are described in more detail in Chap. 3.

Cycles Based on Single-Phase Working Fluids

More theoretical thermodynamic reference cycles and heat pump concepts are based on gaseous working fluids than on working fluids that change phase. Phase-change machines follow only the Clausius-Rankine and Lorenz cycles. Machines operating without phase changes most commonly include Otto-, Diesel-, Brayton-, Ericsson-, and Stirling-cycle concepts.

This book does not include analyses of Brayton, or gas turbine, cycles. (These are referred to as Joule or Ericsson cycles in Europe, but they are not related to the Ericsson cycles described in this book, and the reader should be careful to avoid being misled when consulting European literature.) There are no Brayton-cycle turbines in the size range of interest, and low-cost positive-displacement equipment for Brayton cycles has not been developed. Because of their lower efficiency, simple Brayton-cycle heat pumps offer no performance advantage to compensate for the equipment cost. Brayton-cycle concepts which use regenerative heat exchange could have higher efficiency levels, but with the penalty of higher equipment costs.

Many engineers understand engine concepts based on the Otto, Diesel, and Brayton cycles, since much effort has gone into developing and refining their physical embodiments in cars, trucks, jet engines, and electric power generators. Brayton-cycle refrigerators are widely used in commercial aircraft, where they have easy access to large amounts of fresh, pressurized air. Also, their low weight offsets their inherent low efficiency. As with Clausius-Rankine equipment, a wealth of literature is available for analyzing their performance.

Practical embodiments of Otto- and Diesel-cycle refrigerators have not been conceived.

Most engineers do not understand the design of equipment based on Stirling cycles and the related Ericsson and Vuilleumier cycles because these cycles have not been in the mainstream of practical applications. Although engines based on Stirling cycles have been developed, heat pump and refrigeration applications of these cycles have not been given much attention. Their analysis is also far more complex, as explained in later chapters. However, they have some characteristics that make them potentially desirable alternatives in specific niches. This book focuses on one of those niches: small heat-powered heat pumps.

Stirling-cycle heat pumps have been commercially successful in cryogenic applications, where they have distinct technical advantages over other practical technologies. However, they have not yet been successful in comfort heating and cooling situations, which have temperature ranges of about -20 to $60°C$ (-5 to $140°F$). As a result, comparative analyses of their performance in comfort heating and cooling configurations are not readily available in the literature.

Concepts that integrate the engine and the refrigerator are, by far, the least understood. The integration itself is important. Analysis of only the separate basic cycles would indicate that the combined performance is equivalent. However, some losses necessarily occur when the power and refrigeration cycles are combined, and only an analysis of the actual performance of practical components shows the differences among the concepts.

Of the cycles mentioned above, only the ideal Stirling and Ericsson cycles are equivalent to Carnot cycles in thermodynamic efficiency. (See Chap. 2 for definitions.) However, we have not selected them for analysis solely because they have maximum thermodynamic efficiencies. Such ideality of one cycle is not a good *a priori* reason to exclude other cycles from consideration. Analysis of ideal cycles, such as Stirling, Ericsson, and Carnot, will indicate that their thermodynamic efficiencies are equivalent. However, the losses that necessarily occur in their real embodiments and through interactions of the power and refrigeration cycles lead to significant differences among them. Conditions of operation may differ, and equipment based on nonideal cycles could actually be more effective than that based on ideal cycles. Furthermore, efficiency is not the only important factor. Other factors which are important in practical systems are described in Chap. 4.

In summary, we analyzed integrated heat pump concepts based on Stirling and Ericsson cycles not only because of their thermodynamic ideality, but also because there are equipment concepts based on these cycles that show good practical potential. Yet they are not widely understood and appreciated by the technical community.

References

1. Thomson, W., "The Power Required for Thermodynamic Heating of Building," *Cambridge and Dublin Mathematical Journal*, 124 (November 1853).
2. Vuilleumier, R., "Method and Apparatus for Inducing Heat Changes," U.S. Patent 1,275,507 (1918).
3. Skrotski, Bernhardt G. A., *Basic Thermodynamics: Elements of Energy Systems*, New York: McGraw-Hill, 1963.
4. Reynolds, W. C., and H. C. Perkins, *Engineering Thermodynamics*, New York: McGraw-Hill, 1977.
5. Strong, D. T. G., "New Gas-Fired Heat Pump Could Halve Heating Bills," *Chartered Mechanical Engineer*, MEP Ltd., Bury St. Edmunds, Suffolk 4426/b-79 (June 1979).
6. Lorenz, H., "Beitraege zur Beurteilung von Kuehlmaschinen," *Zeitschrift des Vereines Deutscher Ingenieure*, 38 (1894).
7. Ibrahim, O. M. and S. A. Klein, "Optimum Power of Carnot and Lorenz Cycles," presented at the 1989 ASME meeting, San Diego, November/December 1989.

Chapter 2

Heat Pump Theory

General Heat Pump Operation

To maintain precision in the analysis discussed in later chapters, we must review some elementary descriptions of heat pump cycles and their thermodynamics. Although this chapter includes some very basic descriptions, we do not intend to teach cycle thermodynamics. Our purpose is to define basic terminology and explain the basis for the comparisons developed in subsequent chapters.

The integrated heat pump concepts analyzed in this book are mechanical systems which operate by means of flowing fluids. Absorption cooling and nonmechanical heat pumping techniques such as thermoelectric devices are not covered. Although absorption chillers are, strictly speaking, fluid-compression devices, they operate too differently from mechanical equipment to be usefully analyzed by similar methods.

Throughout this chapter, we distinguish between equipment, equipment concepts, and cycles. We refer to machinery, heat pumps, refrigerators, or engines as *equipment*. *Concepts* are ideas for assembling equipment components into a heat pump. *Cycles* represent the changing thermodynamic conditions of a working fluid as it passes through the equipment.

All mechanical heat pumps have two principal elements: a refrigerator, which carries heat from a lower-temperature environment to a higher-temperature environment, and an engine, which provides the necessary mechanical energy (work) to drive the refrigerator. Refrigerators are still properly called refrigerators even when they are used for heating rather than cooling. Figure 2.1 shows a general mechanical refrigeration cycle. After the working fluid, or refrigerant, is compressed, heat exchanger HX1 removes the heat of compression at a high temperature. The refrigerant may change phase when the heat is

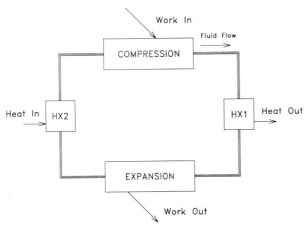

Figure 2.1 General mechanical refrigeration cycle.

removed, or it may remain in one phase. One-phase cycles are usually gas-compression cycles because the thermodynamic properties of liquids and solids do not lead to good equipment performance. After expansion, the refrigerant can absorb heat at a lower temperature in heat exchanger HX2.

In closed cycles, the refrigerant is recirculated. In open cycles, it is not. In some cases it is more economical to use an open cycle with a low-cost refrigerant such as air. This eliminates the need for one heat exchanger, since the air can be discarded to the environment rather than returned to its original state. Systems which use atmospheric air and operate below atmospheric pressure need HX2, but not HX1.

Refrigerators can use many different types of compressors. There are two general classes: positive-displacement (reciprocating-piston and rotary compressors) and dynamic (radial and axial turbocompressors and jet compressors). Operating these compressors requires the input of mechanical work, which leads to the term *mechanical heat pump*. Expansion may be achieved by a flow restriction or by a mechanical compressor operating in a reverse manner as an expander. An expander costs more than a flow restriction, but it can provide mechanical work that may help drive the refrigerant compressor. Experience (as expressed in the second law of thermodynamics) tells us that the expander can only provide a limited amount of work. It can never provide enough work by itself to drive the compressor.

Most refrigeration cycles can be conceptually changed into engine cycles by operating HX1 at a higher temperature so that heat is added to the working fluid. This reverses the direction of heat flow, as shown in Fig. 2.2. The addition of heat at HX1 allows the expander to pro-

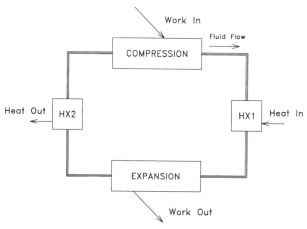

Figure 2.2 General engine cycle.

duce more mechanical work than is needed to drive the compressor. The net effect is that the heat input produces mechanical work. The fluid in an engine, as in a refrigerator, is also called a working fluid, but it is not correctly called a refrigerant.

Many different engine configurations are possible. For example, some internal combustion engines, such as Diesel- and Otto-cycle machines, combine the functions of the compressor and the expander in the same cylinder. They do not need the heat exchangers because the heat comes in as the chemical energy of a fuel and leaves with the products of combustion. Many steam engines operate with open cycles and consist only of heat exchanger HX1, an expander, and a water feed pump. Of course, closed-cycle, condensing steam engines also use heat exchanger HX2.

Figure 2.3 shows one straightforward way to combine an engine cycle and a refrigerator cycle into a heat pump cycle. There are many other ways to combine them and many possible arrangements for transferring mechanical work from the expander to the compressor. Often an energy-storing flywheel stores this work briefly until it is needed. Work can even be transferred from the engine to the refrigerator indirectly by electric motor-generator combinations. The engine and refrigerator can also share heat exchangers to reduce the equipment cost.

In principle, heat pumps can be driven by almost any form of energy, although only a few forms are practical and effective. One important variation on the general engine-driven refrigerators described above is refrigerators with thermally driven compressors. Figure 2.4 illustrates a thermal compression cycle, and two of the six concepts analyzed in later chapters use thermally driven compressors. Al-

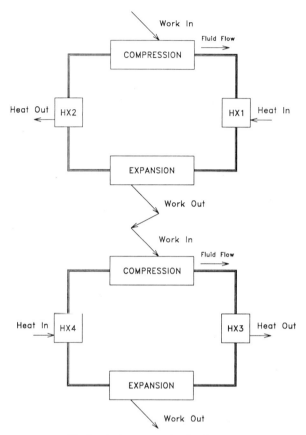

Figure 2.3 Mechanical-compression heat pump.

though it is more difficult to visualize how they operate, these machines have a potential for good performance and practical designs.

Only a few of the many possible heat pump configurations are used commercially. This is because the cost of the equipment is usually more important than the cost of the energy it uses. It is not practical to use a large number of components because this increases equipment cost, and limiting the number of components reduces the number of possible combinations. Most residential and small commercial heat pumps are driven electrically. That is, a large engine at the central power plant indirectly drives many small refrigerators. Most of these systems do not recover the heat from heat exchanger HX2 of the engine at the central power plant, and this reduces overall system efficiency. However, the cost savings from using a single large engine instead of many small ones offsets this disadvantage.

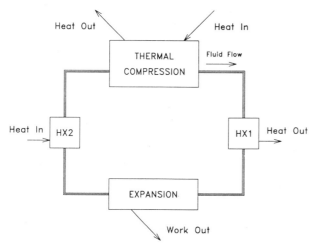

Figure 2.4 Thermal-compression heat pump.

General Categories of Reference Cycles

The thermodynamic cycles that describe the sequential changes in the state of the working fluid as it moves through a heat pump are useful standards of reference during design and evaluation. Three categories of reference cycles are generally recognized.

Theoretical cycles are the series of thermodynamic states of a working fluid as it passes through perfect equipment. Perfect equipment is hypothetical equipment whose components all operate as well as possible within the laws of thermodynamics (fully effective heat exchange and no friction or other irreversible process). The operating cycles of several perfect engines and refrigerators are customarily used as a basis of comparison for the performance of similar real equipment. The performance of these cycles depends on the thermodynamic properties of the working fluids used. Therefore, it is important to identify the working fluid when referring to such a cycle.

Ideal cycles are theoretical engine or refrigerator cycles which have the best possible performance within the constraints of the second law of thermodynamics. The theoretical Carnot cycle, Stirling cycle, and Ericsson cycle are ideal cycles. As refrigeration cycles, they move the most heat possible per unit of mechanical work input. As engine cycles, they provide the most mechanical work possible per unit of heat input. Chapter 3 describes these cycles. For an ideal cycle, the amount of heat moved per unit of mechanical work input depends only on the temperature levels of the two heat exchangers. All ideal cycles have equivalent efficiency, and their efficiency does not depend on the properties of the working fluid. This independence makes them very useful reference standards. The theoretical Brayton and Clausius-Rankine cycles, which do not have isothermal heat-exchange steps, are not ideal cycles. They are inherently less efficient than ideal cycles that operate between the same temperature limits.

Real cycles represent the behavior of refrigerants in actual equipment. Heat exchange may not be complete or isothermal. Compressors and expanders may have leakage, heat loss or gain, dead space, and friction losses. All these factors make real equipment perform less well than the theoretical cycles which approximate them.

The specific reference cycles considered in this book are described in Chap. 3. The following definitions of performance parameters will clarify how theoretical and ideal reference cycles are used in describing and evaluating the performance of real equipment.

Performance Parameters

The performance of heat pumps is characterized by three basic parameters:

- Coefficient of performance, COP (in Europe, this is called refrigerating factor). COP_h refers to the performance of heating equipment, and COP_c refers to cooling-equipment performance.
- Refrigeration (or cooling) efficiency, η.
- Specific volumetric or mass cooling (heating) capacity, q_v or q_m, respectively.

We refer to these parameters throughout our analysis. Three other factors are also important and are used in some segments of the analysis:

- The specific volumetric or mass flow rate of the working fluid, which characterizes the size of the equipment and the potential level of flow friction losses. It is primarily used in Chap. 8.
- Engine (prime mover) efficiency, which, for integrated heat pumps, is an implicit component of performance calculations.
- Economic factors, which are mentioned further in Chap. 4. These factors are of primary importance from a practical standpoint, but

they have only general relevance to the technical analysis of performance in this book. Therefore, they are not covered in detail.

Coefficient of performance

The coefficient of performance (COP) of a refrigerator is defined differently for heating and cooling applications. For heating applications, it is the ratio of the heat delivered at the high-temperature heat exchanger to the net energy input driving the refrigeration cycle. For mechanical refrigerators, the energy input is the net work input to the compressor. For cooling applications, it is the ratio of the heat absorbed at the low-temperature heat exchanger to the net energy input. The energy units used in the numerator and denominator are the same, so COPs are dimensionless and without units.

The ideal COP is the COP of a thermodynamic cycle that operates reversibly between the source and sink temperatures in such a way that it rejects and absorbs the heat isothermally. Although this definition of ideal COP is usually derived for a Carnot cycle, the Carnot-cycle COP is identical to the COPs of all other ideal cycles operating between the same source and sink temperatures. Therefore, it is more comprehensive to use the term *ideal COP* than to refer to the COP of a Carnot cycle.

The COP of an ideal cycle can be calculated from the input and output temperature levels only. The COP of a theoretical cycle which is not ideal can be calculated from the cycle thermodynamics. The COPs of real cycles and equipment can be approximated by more detailed calculations which simulate its operation, but the true test of a real COP is measurement. Calculated COPs should always be clearly identified as to whether they are for ideal cycles, theoretical cycles, or real cycles.

For an ideal heating cycle, the above energy ratios that define the COP can be reduced to a simple dimensionless relationship between the temperatures of the heat source and heat sink:

$$COP_{h,i} = \frac{\text{absolute temperature of the sink}}{\text{sink temperature - source temperature}}$$

For an ideal cooling cycle, the corresponding COP is

$$COP_{c,i} = \frac{\text{absolute temperature of the source}}{\text{sink temperature - source temperature}} = COP_{h,i} - 1$$

The above two expressions hold true for all refrigerants. The ideal COP is the maximum thermodynamic limit for the COP between two temperature levels. In principle, there are infinitely many theoretical cycles which are ideal. However, there are also infinitely many theoretical cycles which are not ideal. The COP of the latter will always be less than the COP of an ideal cycle which operates between the same temperature levels. The COP of a real cycle will never be greater than the COP of its corresponding theoretical cycle because a real cycle will always have some irreversibility.

The COP of a theoretical heating cycle is

$$\text{COP}_{h,th} = \frac{\text{heat delivered to a high-temperature sink}}{\text{net energy input to drive the cycle}}$$

The heating COP of a real refrigerator is

$$\text{COP}_{h,r} = \frac{\text{actual heat delivered to a high-temperature sink}}{\text{actual net energy input to drive the cycle}}$$

The above two equations have similar counterparts for cooling equipment, based on heat removed from a low-temperature source.

Refrigeration efficiency

The refrigeration efficiency (η) is the ratio of two COPs. The denominator is the COP of the reference cycle. It is important to indicate (usually by subscripts) whether the COPs are ideal, theoretical, or real. For example,

$$\eta_{th,i} = \frac{\text{COP}_{th}}{\text{COP}_i}$$

$$\eta_{r,i} = \frac{\text{COP}_r}{\text{COP}_i}$$

$$\eta_{r,th} = \frac{\text{COP}_r}{\text{COP}_{th}}$$

Of course, we do not mix heating and cooling COPs when calculating refrigeration efficiency ratios.

The first of these ratios, $\eta_{th,i}$, is usually used when evaluating the performance of different working fluids in Clausius-Rankine cycles because such evaluations are usually concerned with the fluid properties and not the specific characteristics of the equipment. For Brayton cycles, the values of $\eta_{th,i}$ can vary from 0 to 1 regardless of the actual sink and source temperatures. This is because the pressure ratio of the theoretical Brayton cycle determines its COP.

The other two refrigeration efficiency ratios, $\eta_{r,i}$ and $\eta_{r,th}$, are more useful for evaluating the performance of equipment or equipment designs.

Specific Volumetric Heating and Cooling Capacity

The size of the mechanical components is an important factor in evaluating any heat pump. Later chapters use the volumetric heating or cooling capacity q_v, as a measure of equipment size. For a heating application,

$$q_{v,h} = \frac{\text{heat delivered to the high-temperature sink}}{\text{volume of the operating working fluid}}$$

and for a cooling application,

$$q_{v,c} = \frac{\text{heat absorbed from the low-temperature source}}{\text{volume of the operating working fluid}}$$

These parameters are usually expressed in units such as kilowatts of cooling per cubic meter of working fluid or tons of refrigeration per cubic foot. In practice, there is a lack of uniformity stemming from differences in reference volumes. Sometimes the total volume of the working space or total displacement of the refrigerator is used. In other cases, volumetric flow rates are expressed as displaced fluid volumes or are defined relative to compressor suction conditions. In later chapters, we use the convenient basis of the total volume of the operating space.

The performance parameters described in this chapter are used to rate heat pump performance. Whenever that is done, one must be very careful in checking the definitions, the units used, and, perhaps most importantly, whether the parameters really give a fair comparison of the performance of very different concepts.

Chapter

3

Reference-Cycle Thermodynamics

This chapter describes the thermodynamic cycles which are used as references for rating heat pump performance. These cycles include the ideal Carnot, Stirling, and Ericsson cycles; the theoretical Lorenz, Clausius-Rankine, and Brayton cycles; and combinations of ideal cycles, including the Vuilleumier.

Detailed thermodynamic analysis of heat pumps, particularly those based on Stirling cycles, is complex. However, insight into heat pump operating characteristics can be obtained with a simple analysis based on performance parameters like those described in Chap. 2 and on the thermodynamics of certain reference cycles. Important theoretical cycles are described in this chapter. These cycles are often used as reference cycles when analyzing heat pump performance. Scientists developed these cycles during the evolution of thermodynamics, and engineers have adopted a system which uses them as consistent frames of reference when calculating the performance of real heat pumps. Reference cycles give us a way to account for the theoretical thermodynamic limits inherent in engine and refrigeration cycles.

Several cycles are described below. They represent perfect performance of certain real equipment classes. We describe the cycles by following the thermodynamic processes in temperature-entropy diagrams, where energy flows can be clearly illustrated and easily followed. We also derive the equations used to calculate the cycle coefficient of performance, and we describe some example embodiments. However, these embodiments are not unique. Each of the cycles can apply to many different equipment configurations, making the design of heat pumps a complex, interesting field. We follow the custom of heat pump engineering and designate the direction of heat and work transfer by subscripts, rather than by sign. All terms in the equations used in this chapter represent positive numerical values, and all temperatures are absolute.

Ideal Reference Cycles

Three common theoretical reference cycles—Carnot, Stirling, and Ericsson—are ideal. That is, they are as thermodynamically efficient as any cycle can be when operating with any specific temperature limits. Ideal engine and refrigerator cycles can be combined to form complete heat pump cycles.

Carnot cycle

In 1824, Nicolas Leonard Sadi Carnot suggested an ideal cycle consisting of isothermal and adiabatic steps. This cycle can be used to represent either engines or refrigerators. Figure 3.1 shows the temperature-entropy diagram of the Carnot refrigeration cycle and an embodiment based on turbocompressors and turboexpanders. This embodiment illustrates the nature of the cycle, although actual operation of this equipment would require an impractical degree of control over the heating and cooling processes.

Consistent with the general description of engine and refrigeration cycles in Chap. 2, the individual steps for the Carnot refrigeration cycle are:

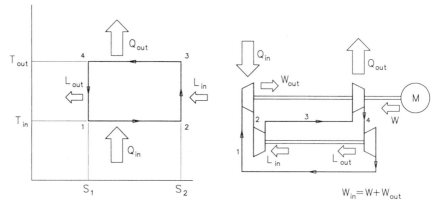

a. Temperature–Entropy Diagram

b. Continuous Flow Embodiment

Figure 3.1 Carnot refrigeration cycle.

1 to 2 Isothermal expansion at low source temperature T_{in}, with heat input Q_{in} equal to work output W_{out}
This work output is transferred to the isothermal compression step to offset part of the work input W_{in} required by that step. For an ideal-gas working fluid, $Q_{in} = W_{out}$ exactly for this isothermal expansion. The gases usually considered as working fluids for gas cycles—air, hydrogen, and helium—have near-ideal behavior, so this equality is a good approximation.

2 to 3 Reversible adiabatic (isentropic) compression requiring mechanical work input L_{in}
This work is exactly equal to the mechanical output L_{out} of the isentropic expansion step because the two steps operate between identical temperature levels.

3 to 4 Isothermal compression at high sink temperature T_{out}, with heat output Q_{out} equal to work input W_{in}
This work input is partly offset by the work output of the isothermal expansion step. The difference, $W_{in} - W_{out}$, provided by a prime mover such as a motor M, is the work required to drive the cycle.

4 to 1 Isentropic expansion, producing mechanical work output L_{out} equal to L_{in}

As shown in Fig. 3.1a, the sequence of these steps is counterclockwise on the temperature-entropy diagram, corresponding to a power-consuming refrigeration cycle. The net cycle energy input in the form of mechanical work W is the difference in the amount of work associated with the isothermal steps:

$$W = W_{in} - W_{out} = Q_{out} - Q_{in}$$

The useful heating and cooling capacities are

$$Q_{out} = Q_{in} + W = T_{out}(S_2 - S_1)$$

$$Q_{in} = Q_{out} - W = T_{in}(S_2 - S_1)$$

The corresponding ideal coefficients of performance are

$$\text{COP}_{h,i} = \frac{Q_{out}}{W} = \frac{T_{out}}{T_{out} - T_{in}}$$

$$\text{COP}_{c,i} = \frac{Q_{in}}{W} = \frac{T_{in}}{T_{out} - T_{in}}$$

For an engine cycle, the sequence would be clockwise, the isothermal expansion would be at the high source temperature, and the isothermal compression would be at the low sink temperature.

Figure 3.1b shows one embodiment of the Carnot cycle—two turbocompressors and two turboexpanders, with net power requirements provided by a prime mover, such as an electric motor M. This simple diagram does not show the complexity of the actual equipment that would be required. Turbomachinery is inherently adiabatic. To achieve the isothermal steps would require precisely controlled removal and addition of heat continuously during the compression and expansion, respectively. This could be approximated by multistage turbines with interstage heat transfer. However, for small units, the cost of multistage compressors with interstage heat exchangers usually far outweighs the value of the energy saved.

This embodiment involves continuous flow of the working fluid. Each infinitesimal element of the working fluid is at a different thermodynamic state, and the state of each element changes smoothly as the element moves through the cycle. Carnot cycles are often described as discontinuous processes, changing the state of the working fluid in intermittent steps. This important distinction between continuous and discontinuous flow has been described well by Hougen and Watson.[1] It does not affect the ideal analysis of cycle performance, but it strongly influences the way in which more realistic analyses are carried out. These analyses are usually far more difficult for discontinuous-flow embodiments. This point is particularly significant for Stirling- and Ericsson-cycle embodiments, which are usually discontinuous. Although the Carnot cycle is often used in thermodynamics texts as the example of an ideal cycle, it is not practical. Besides the difficulty of attaining isothermal operation, its specific heating and cooling capacities are relatively low. Both Stirling and Ericsson ideal cycles offer greater refrigeration capacity.

An interesting open-cycle concept which follows one-half of the Carnot cycle in operation is the Kelvin refrigerator,[2,3] shown in Fig. 3.2. This diagram also does not show the practical difficulty of achieving the isothermal step.

Stirling and Ericsson cycles

Two other ideal reference cycles are named after Robert Stirling and John Ericsson, who conceived and designed different types of hot-air engines in the first half of the nineteenth century. These designs later led to the formulation of the ideal Stirling and Ericsson cycles. Stirling and Ericsson used embodiments that change the state of the working fluid in intermittent steps, as described in detail in Chap. 5. However, continuous-flow embodiments are shown in the following figures for consistent comparison with the other cycles described in this chapter.

Figure 3.3 shows the temperature-entropy diagram for the Ericsson refrigeration cycle and a continuous-flow embodiment. The sequence of steps is:

Reference-Cycle Thermodynamics   23

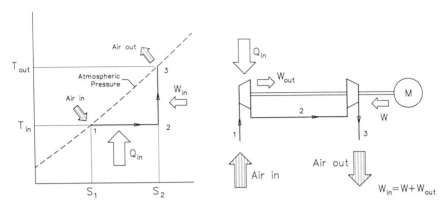

a. Temperature–
   Entropy Diagram

b. Continuous Flow
   Embodiment

Figure 3.2 Kelvin heat pump.

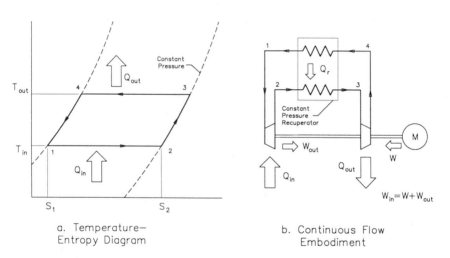

a. Temperature–
   Entropy Diagram

b. Continuous Flow
   Embodiment

Figure 3.3 Ericsson refrigeration cycle.

1 to 2	Isothermal expansion at low source temperature T_{in}, with heat input Q_{in} equal to work output W_{out}
2 to 3	Isobaric (constant-pressure) heat exchange, with heat input Q_r supplied by a recuperator in continuous-flow embodiments and a regenerator in discontinuous-flow embodiments
3 to 4	Isothermal compression at high sink temperature T_{out}, with heat output Q_{out} equal to work input W_{in}
4 to 1	Isobaric heat exchange, with heat output Q_r transferred through a recuperator in continuous-flow embodiments and a regenerator in discontinuous-flow embodiments

The continuous-flow mechanization of the Ericsson cycle, shown in Fig. 3.3b, is similar to that of the Carnot cycle. The isobaric heat-exchange steps (2 to 3 and 4 to 1) in the Ericsson cycle can be achieved through a recuperative heat exchanger. However, as with the Carnot cycle, this embodiment would have difficulty achieving isothermal operation of the turbocompressor and turboexpander.

Figure 3.4 shows the temperature-entropy diagram for the Stirling refrigeration cycle and a continuous-flow embodiment. The sequence of steps is:

1 to 2   Isothermal expansion at low source temperature T_{in}, with heat input Q_{in} equivalent to work output W_{out}

2 to 3   Isochoric (constant-volume) compression, with heat input Q_r supplied by a regenerator

3 to 4   Isothermal compression at high sink temperature T_{out}, with heat output Q_{out} equivalent to work input W_{in}

4 to 1   Isochoric expansion, with heat output Q_r stored in a regenerator

A continuous-flow mechanization of the Stirling cycle is more difficult than with the Carnot and Ericsson cycles. As with the Carnot and Ericsson cycles, the Stirling embodiment shown in Fig. 3.4b would have difficulty achieving isothermal operation of the turbocompressor and turboexpander. However, Fig. 3.4b includes another component whose construction seems impossible for continuous-flow equipment. A recuperative heat exchanger that can achieve compression or expansion of a fluid at constant specific volume has not yet been invented. Regenerative heat exchange is also not possible because the regenerator would have to move from one flow stream to the other,

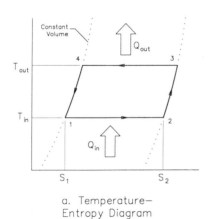

a. Temperature–Entropy Diagram

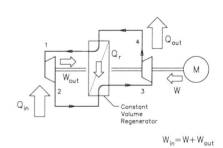

b. Continuous Flow Embodiment

Figure 3.4 Stirling refrigeration cycle.

which seems impossible because the two streams have varying pressures.

Both Stirling and Ericsson cycles are more adaptable to discontinuous-flow embodiments, and regenerative heat exchange is critical to these embodiments. Fixed regenerators can be incorporated into the working space of discontinuous-flow Stirling-cycle machines straightforwardly. Adaptation of regenerators to the isobaric steps of discontinuous-flow Ericsson-cycle machines requires the use of valves to maintain constant pressure.

For the same temperature ratio and the same specific volumetric change of the working fluid, the Stirling and Ericsson refrigeration cycles have greater cooling capacity than the Carnot cycle, as shown in Fig. 3.5. Similarly, Stirling- and Ericsson-cycle engines have greater power output than an equivalent Carnot-cycle engine. Carnot-cycle embodiments are generally considered impractical because of the difficulty of achieving isentropic expansion and compression and because of their smaller power output at the same operating conditions. The pressure difference in a Stirling cycle is the same as in a Carnot cycle with the same volumetric change and temperature levels. However, an Ericsson cycle with the same pressure ratio has a higher volumetric change, which gives the Ericsson cycle a theoretical capacity advantage.

Other Theoretical Reference Cycles

Three common theoretical reference cycles—Lorenz, Clausius-Rankine, and Brayton—are described in this section for comparison with the ideal cycles just described. Although they are not ideal cy-

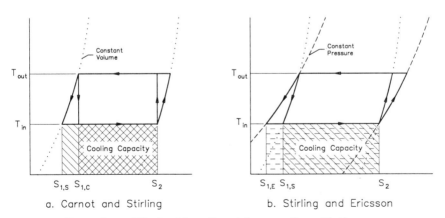

Figure 3.5 Comparison of ideal refrigeration cycles operating with the same pressure ratio and between the same temperatures.

cles, they are useful because they approximate the operation of real machines that are in wide application.

Lorenz theoretical cycle

The Lorenz cycle was first suggested for evaluating real processes that have heat exchangers in which the temperature of the working fluid changes as a result of finite heat capacity. This variation on the Carnot cycle, shown in Fig. 3.6, was first described by H. Lorenz in 1894.[4] The temperature-entropy diagram shows that the adiabatic steps in the Lorenz cycle, 1-2-3-4, are similar to Carnot-cycle steps 1-2'-3-4'. However, the isothermal compression and expansion steps of the Carnot cycle, which involve heat transfer, are modified to occur at varying temperatures which are functions of entropy. This represents heat transfer to a heat sink and from a heat source, both of which have finite heat capacity. In one limiting case, these two heat-transfer steps represent the constant-pressure phase change of a two-component refrigerant mixture. An embodiment of this cycle can be visualized to be the same as that shown for the Carnot cycle in Fig. 3.1b, the only difference being in the heat-exchanger design.

An ambiguity arises in calculating refrigeration efficiencies which involve the COP of a corresponding ideal reference cycle. Lorenz cycles, unlike ideal reference cycles, do not have single source and sink temperatures. Therefore, the Lorenz-cycle COP cannot be represented by two temperatures that are unambiguously related to an ideal reference cycle, and the temperatures used in calculating refrigeration efficiency must be specified carefully. Applying thermodynamic relationships similar to those used for the Carnot cycle, the heating coef-

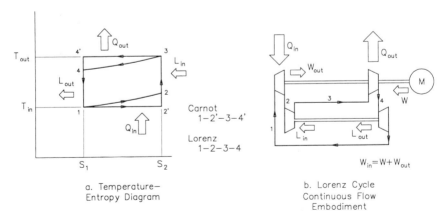

Figure 3.6 Comparison of the Lorenz and Carnot refrigeration cycles.

ficient of performance for a Lorenz refrigeration cycle in which temperature is a linear function of entropy is

$$\text{COP}_{h,t} = \frac{(T_3 - T_4)/\ln(T_3/T_4)}{(T_3 - T_4)/ln(T_3/T_4) - (T_1 - T_2)/\ln(T_1/T_2)}$$

In the above equation, log-mean average temperature differences characterize the heat-exchange steps. The cooling-cycle COP has a comparable form.

Clausius-Rankine theoretical cycle

Most commercial heat pumps, air conditioners, and refrigerators that are currently marketed are based on vapor-compression cycles. The reference cycle that most closely approximates their operation is a theoretical cycle named after two renowned physicists, Rudolf J. E. Clausius (Germany) and W. J. M. Rankine (Scotland). Its first embodiment was a closed-cycle ice-making machine that used ethyl ether as a refrigerant. Jacob Perkins patented this machine in 1834 and built it in England about 1835.[5]

From Fig. 3.7, it is evident that this reference cycle is not ideal. The temperature varies during heat-rejection steps 2 to 4, and expansion step 4 to 5 is an irreversible throttling. Yet, heat pumps operating on this principle are highly practical, offering relatively low cost, high COP, and high specific heating and cooling capacities. The COP of Clausius-Rankine cycles cannot be derived simply in terms of absolute temperatures. To account for the latent heats of phase changes, the expression for COP must use enthalpies:

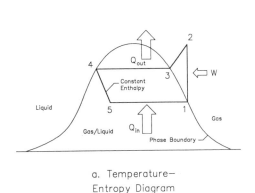
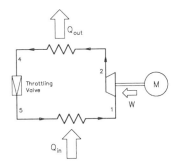

a. Temperature–Entropy Diagram

b. Continuous Flow Embodiment

Figure 3.7 Clausius-Rankine refrigeration cycle.

$$\text{COP}_{h,t} = \frac{h_2 - h_4}{h_2 - h_1}$$

Brayton theoretical cycle

The Brayton reference refrigeration cycle is a special case of a Lorenz cycle with constant-pressure heat-transfer steps. An ice machine which followed an open Brayton cycle with air as a refrigerant was first built by John Gorrie, a doctor in Apalachicola, Florida, in the 1840s. His British patent is dated 1850; his U.S. patent, 1851.[6] However, Gorrie's contribution was forgotten, and in 1873 the concept was independently invented again by two men, George Brayton and James Joule. George Brayton, a Boston engineer, used principles learned in the development of his heat engine to design a refrigerator that combined constant-pressure and constant-entropy steps. At the same time, an English brewer, James Joule, proposed equipment which follows the same cycle. Because of this parallel invention path, this cycle is called Brayton in the United States and Joule or Ericsson in Europe. This cycle is not the Ericsson cycle described elsewhere in this book, and the reader should take care to avoid confusion over this terminology when consulting European literature. In the Brayton cycle, the working fluid always remains in the gas phase. Both Joule's and Brayton's embodiments used air as the working fluid, and this cycle is often referred to as an air cycle, regardless of the actual working fluid used.

The Brayton-cycle temperature-entropy diagram in Fig. 3.8a shows that no constant-temperature steps occur. Therefore, the cycle COP

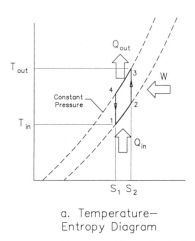
a. Temperature–
Entropy Diagram

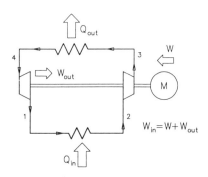
b. Continuous Flow
Embodiment

Figure 3.8 Brayton refrigeration cycle.

cannot be expressed by two temperatures that are unambiguously related to an ideal reference cycle. As is the case with Lorenz cycles, refrigeration efficiency calculations must be handled carefully. The COP of this cycle is expressed in terms of the energy associated with the working fluid (total mass m), as it passes through the individual steps. The cycle work input is

$$W = mC_p[(T_3 - T_2) - (T_4 - T_1)]$$

and the useful heat output is

$$Q_{out} = mC_p(T_3 - T_4)$$

Therefore, the heating COP for this cycle is

$$COP_{h,t} = \frac{Q_{out}}{W} = \frac{(T_3 - T_4)}{(T_3 - T_2) - (T_4 - T_1)}$$

When expressed in terms of the pressures associated with the isentropic expansion steps, this equation becomes

$$COP_{h,t} = \frac{1}{1 - (P_2/P_3)^{(\gamma-1)/\gamma}}$$

and the cooling COP is

$$COP_{c,t} = \frac{1}{(P_3/P_2)^{(\gamma-1)/\gamma} - 1}$$

In these equations, $\gamma = C_p/C_v$, C_p is the specific heat of the refrigerant at constant pressure, and C_v is the specific heat at constant volume. The derivation of the above expressions assumes that γ is constant throughout the cycle.

Combinations of Cycles

Two types of combined cycles have practical significance. One combines refrigeration cycles (counterclockwise on the temperature-entropy diagram). These cycles operate in stages to increase the temperature of the heat delivered by the previous stage. In the second type, a power-producing (clockwise) cycle drives a refrigeration cycle. Both combinations require an energy input; for the first type it is

Staged refrigerators

This type of combination requires external work to drive each cycle. A practical example of staged refrigerators is the "desuperheater heat pump." This refrigerator raises the temperature level of heat that would normally be transferred to the outdoors from an air conditioner and uses it for another purpose, such as water heating.

Figure 3.9 represents a two-stage refrigerator. For ideal cycles, the net effect of this combination of cycles is the same as if they were a single cycle operating between the highest and lowest temperature levels. The delivered heat output and the overall COP are not affected by this cycle staging. Therefore, its ideal reference heating COP is

$$\text{COP}_{h,i} = \frac{T_h}{T_h - T_c}$$

Combined engine-refrigerator cycles

Figure 3.10 shows temperature-entropy diagrams for the second of the above combinations. As can be seen from the figure, this combination of an ideal refrigeration cycle and an ideal engine cycle has three characteristic temperatures which define its COP. Equating the entropy differences from Fig. 3.10 leads to

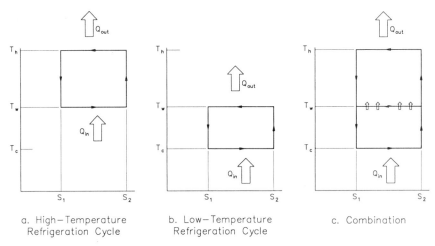

a. High–Temperature Refrigeration Cycle

b. Low–Temperature Refrigeration Cycle

c. Combination

Figure 3.9 Combination of refrigeration cycles.

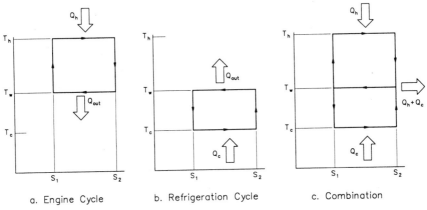

a. Engine Cycle **b. Refrigeration Cycle** **c. Combination**

Figure 3.10 Temperature-entropy diagram for an ideal heat-driven heat pump.

$$\frac{Q_h}{T_h} + \frac{Q_c}{T_c} = \frac{Q_h + Q_c}{T_w}$$

Algebraic rearrangement of this equation leads to

$$Q_c = \frac{Q_h(T_h - T_w)T_c}{T_h(T_w - T_c)}$$

which can be further used to express the ideal heating COP:

$$\text{COP}_{h,i} = \frac{Q_h + Q_c}{Q_h} = \frac{T_w(T_h - T_c)}{T_h(T_w - T_c)}$$

This COP is qualitatively different from the COPs discussed previously because it includes the ideal engine's efficiency. The above equation applies whether the two individual ideal reference cycles are different or the same, or whether they are Carnot, Stirling, or Ericsson, as long as the heat-transfer steps take place isothermally at the same three temperatures.

Vuilleumier cycle

A practical variation of an integrated engine-refrigerator combination is named after its inventor, Rudolph Vuilleumier, who patented his first refrigeration machine in 1917. This cycle and its classical configuration, shown in Fig. 3.11, closely resemble the ideas Vuilleumier expressed in his 1918 U.S. patent.[7]

No continuous-flow embodiment of this cycle has been conceived. Therefore, unlike the other cycle descriptions in this chapter, this de-

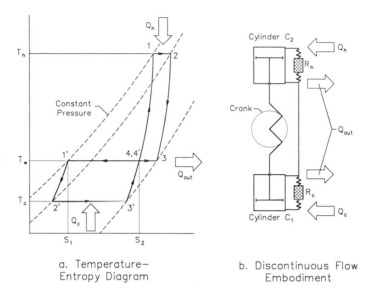

a. Temperature–
Entropy Diagram

b. Discontinuous Flow
Embodiment

Figure 3.11 Ideal Vuilleumier cycle.

scription uses a discontinuous-flow embodiment, shown in Fig. 3.11. The overall process combines the effects of a power-producing cycle, 1-2-3-4, acting directly on a refrigeration cycle, 1'-2'-3'-4'. The two cycles are combined between the common pressure levels of 1-1', 2-2', and 3-3', indicated by dashed lines. The sequence of steps is:

Vuilleumier Cycle—Engine Segment (Cylinder C_2)

1 to 2   Heat Q_h enters the working fluid through a heat exchanger at temperature T_h.

2 to 3   The fluid isochorically releases its heat to regenerator R_h.

3 to 4   The useful heat is released through a heat exchanger at temperature T_w.

4 to 1   The working fluid isochorically absorbs the heat previously stored in regenerator R_h.

Vuilleumier Cycle—Refrigerator Segment (Cylinder C_1)

1' to 2'   The high-pressure working fluid flows through regenerator R_c, isochorically rejecting heat to the regenerator without power consumption.

2' to 3'   Heat Q_c at low temperature T_c is continuously absorbed through a heat exchanger while pressure decreases from P_2' (which is identical to P_2) to P_3'.

3' to 4'   The high-pressure working fluid flows through regenerator R_c, isochorically absorbing heat from the regenerator without power consumption.

4' to 1'   Heat is rejected through a heat exchanger at temperature T_w while the working fluid is compressed from the low-pressure state, P_4, to the maximum pressure, P_1.

In this cycle, heat Q_c supplied at temperature T_c will be pumped to a higher temperature T_w, and useful heat Q_{out} from both the engine and the refrigerator segments is rejected. The areas within 1-2-3-4 and 1'-2'-3'-4' are the same. The work delivered by the engine segment is equal to the work absorbed by the refrigerator.

Some insight into this cycle's performance potential can be gained by inspecting the relationship between COP and temperature. Figure 3.12 shows $COP_{h,i}$ as a function of the low temperature T_c for T_w = 60°C (140°F) and several values of T_h. The small influence of T_c on $COP_{h,i}$ for this cycle represents an important advantage over Clausius-Rankine vapor-compression machines in cold climates.

Summary

Stirling and Ericsson cycles offer the prospect of ideal-cycle performance in practical equipment. For example, Stirling engines and Stirling refrigerators can be combined to form Stirling-Stirling heat pumps. A related cycle, the Vuilleumier, combines the engine and refrigerator in a single device which has three characteristic operating temperatures. As with the Stirling and Ericsson engine cycles, heat input at a high temperature drives this cycle. The refrigerator segment of a Vuilleumier heat pump accepts heat at a low temperature. Rejected heat from both segments flows from a part of the machine

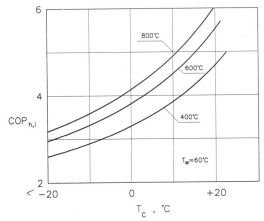

Figure 3.12 Ideal heating COP for a three-temperature, combined engine-refrigerator cycle heat pump.

that is common to both segments. This rejected heat is the Vuilleumier heat pump's delivered energy.

References

1. Hougen, Olaf A., and Kenneth M. Watson, *Chemical Process Principles—Part Two: Thermodynamics,* chap. 13, "Expansion and Compression of Fluids," New York: John Wiley & Sons, Inc., 1947.
2. Thomson, W., "The Power Required for Thermodynamic Heating of Building," *Cambridge and Dublin Mathematical Journal,* 124 (November 1853).
3. Wurm, J., "Advanced Heat Pump Options," pp. 243–245, paper presented at *Future Alternatives in Residential and Commercial Space Conditioning,* sponsored by Institute of Gas Technology, Chicago, June 12-14, 1980.
4. Lorenz, H., "Beiträge zur Beurteilung von Kuehlmaschinen," *Zeitschrift des Vereines Deutscher Ingenieure,* 28(3): 62–68 (1894).
5. Thévenot, R., *A History of Refrigeration Throughout the World,* Paris: International Institute of Refrigeration, 1979, p. 448.
6. Gorrie, J., "Improved Process for the Artificial Production of Ice," U.S. Patent No. 8080 (May 6, 1851).
7. Vuilleumier, R., "Method and Apparatus for Inducing Heat Changes," U.S. Patent 1,275,507 (1918).

Chapter 4

Overview of Integrated Heat Pump Concepts

This chapter describes the factors that influence heat pump design and indicates which factors are most important. We show how the inherent characteristics of integrated heat pumps help to satisfy these important factors. Six concepts are selected as examples of integrated heat pumps. We describe how these concepts are related to one another and to other heat pump concepts, and we present a brief history of the research and development (R&D) that led to their present status.

Desirable Attributes of Heat Pumps

Heat pumps can achieve higher heating efficiencies than combustion equipment, but higher efficiency is pointless if the equipment to achieve it is not affordable. The main dilemma in defining a better engine-driven heat pump configuration is the balance between high efficiency and low cost. In practice, striking this balance is difficult because it is not clear how consumers actually balance equipment cost against future energy cost savings. However, the cost of the equipment is usually the dominant factor. A heat pump which costs significantly more than other space-conditioning devices, no matter how great its efficiency, may not be commercially successful.

Other practical factors are also important in selecting heat pump concepts. The following list indicates factors that have been considered important to the commercial success of any new air-conditioning or heat pump concept:

- Cost of the equipment and its installation
- Operating costs

- Dependability
- Environmental acceptability and safety of the working fluids used
- Ability to maintain reasonable capacity in severe weather
- Ability to deliver heat at high enough temperatures for comfort heating
- Reasonable size and weight
- Ease of manufacture in mass production
- Ease of field servicing using widely known methods
- Adaptability to all heat delivery media
- Ability to drive its own air-handling fans and other auxiliaries

To meet all these criteria is difficult. Yet, some integrated heat pump concepts have been advanced to where they meet many of them. Our reasons for introducing their comparative analysis to technical audiences are more than academic. We are not only filling an information gap; we are guiding the reader toward considering similar new heat pump concepts.

Classification of Heat-Activated Heat Pumps

There are many combinations of thermodynamic effects that can be used for heat pumps, but only a few have been seriously considered for comfort heating and cooling. The strong need for low equipment cost and for high efficiency and ease of manufacture using common technologies has kept R&D focused on the most familiar possibilities. Figure 4.1 represents the concepts that have been commercialized (although not necessarily for comfort heating or cooling) or have received significant R&D attention as heat pumps. Except for the absorption and thermoelectric concepts, they all can be categorized as mechanical heat pumps, fitting the general heat pump model described in Chap. 2. This book is only concerned with concepts in the Vuilleumier, Ericsson-Ericsson, and Stirling-Stirling categories at the top of Fig. 4.1.

The engine and refrigerator segments of most mechanically linked heat pumps can be clearly distinguished. Usually, the separate engine and refrigerator are connected by a shaft or other mechanical linkage. For classification purposes, these split-cycle machines are usually given a compound name, such as Stirling-Rankine, denoting a Stirling-cycle engine driving a Clausius-Rankine-cycle refrigerator.

Vuilleumier-cycle heat pumps do not fit this nomenclature because there is no distinct demarcation between the engine and refrigerator

Overview of Integrated Heat Pump Concepts 37

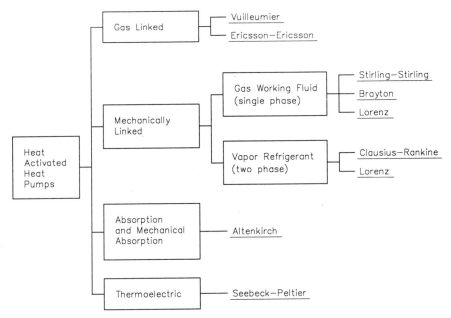

Figure 4.1 Heat-activated heat pumps.

segments. Some portions of the equipment are common to both the engine function and the refrigerator function. The engine and refrigerator also share working fluid. The latter feature distinguishes Vuilleumier heat pumps from the other mechanical heat pump concepts. However, there are other heat pump concepts in which the engine and refrigerator can be clearly distinguished, yet their connection is more intimate than in split-cycle machines. We also classified these other machines as integrated heat pumps, and we selected two Stirling-Stirling concepts as examples.

The simple theoretical classification scheme described above is more complex in practice. It is not always clear which theoretical thermodynamic cycle best describes a particular engine or refrigerator and should therefore serve as its reference cycle. The inventor may also assign a theoretical cycle name to a new device that others would classify differently. Therefore, the names of heat pump concepts should not always be taken literally. One of the six concepts analyzed in this book provides a good example of this. The inventor named it Ericsson-Ericsson, because both segments approximately follow Ericsson cycles. However, the engine and refrigerator segments share part of the working space, and the working fluid moves freely between the two segments. Because of this mingling of fluid and function, this concept is classed as Vuilleumier, but in deference to the inventor, we retained the Ericsson-Ericsson name.

Advantages of Integrated Heat Pumps

Past development of heat-powered heat pumps has been more practical than academic. The developers have concentrated on the practical factors which are important to the commercial success of any new air-conditioning or heat pump concept. Although this book covers equipment concepts that require more sophisticated theoretical treatment than other heat pumps, it does not neglect practical aspects.

Good heat pump performance leads to the energy cost advantages that can make heat pumps an attractive alternative to direct heating. This is especially true for heat pumps based on regenerative cycles, such as Stirling, Ericsson, and Vuilleumier. However, low equipment cost is still the main design criterion. We intentionally selected concepts which are relatively simple, because that is the key to low equipment cost. Another important factor is equipment life and reliability. At least some of the concepts described below have moderate operating conditions that do not unduly tax the mechanical heat pump components. The main attributes of these heat pump concepts are:

- They can be hermetically sealed.
- The heat exchangers which reject heat from the engine and the refrigerator either are the same or are of the same type and are located close together. This can reduce heat-exchanger cost, and it allows the exhaust heat from the engine to be used directly for space heating.
- The working fluids do not change phase. Therefore, the equipment will have broader ranges of temperature service than Clausius-Rankine-cycle refrigerators.
- The working fluids (usually air or helium) are not toxic, corrosive, or environmentally harmful, and their production is also benign.
- The engine and the refrigerator use the same working fluid. In some concepts, the fluid flows back and forth between engine and refrigerator, transferring power from one to the other without mechanical coupling through shafts or other linkages. This can save both cost and energy and can increase reliability.
- Some of the heat pump concepts use displacers rather than pistons. Displacers move fluid at low pressure differences, and they need not be as leak-free as pistons. They have less sliding friction and less stringent lubrication needs.

- Some of these concepts do not need a motor to start or run.
- The external combustion engine segments do not require closely timed ignition and tolerate a range of fuel compositions.
- Some of these concepts can also produce their own auxiliary electric power.
- High theoretical efficiencies give the designer opportunities to trade performance for lower equipment cost.

Selection of Concepts for Analysis

We selected concepts for the analysis described in Chap. 7 to provide examples that would show the wide range of possible conceptual embodiments and the good performance potential of integrated heat pumps. We chose concepts that have high theoretical effectiveness, simple embodiments, and ease of control and capacity modulation. Because the Vuilleumier cycle matches these criteria very well, we selected four different Vuilleumier embodiments:

- The classic single-cycle, two-displacer design, referred to as the traditional Vuilleumier heat pump.
- A concept similar to the traditional Vuilleumier heat pump that uses rotating displacers and internal heat exchangers, called the Vuilleumier heat pump with internal heat exchangers.
- A free-piston Vuilleumier concept, referred to as the Ericsson-Ericsson heat pump by its inventor.
- A four-cycle concept called the balanced-compounded Vuilleumier heat pump by its inventor. Multiple-cycle machines have the advantage of reducing the size of the flywheel or similar massive component used for temporarily storing energy during the cycle. The flywheel mass adds to equipment cost and weight, and can cause vibration.

One criterion that others have used as a basis for classifying some heat pumps as Vuilleumier-cycle is whether their working space has a constant total volume. However, recent inventions[1,2] have established that variable-volume engine and refrigerator segments can also fully share working fluid. The last two of the four Vuilleumier concepts listed above have this characteristic. As Chaps. 5 and 7 show, variable working-space volume allows the machine to achieve both thermal and mechanical compression of the working fluid, leading to higher specific volumetric refrigeration capacity.

Other closely related integrated heat pump concepts have performance advantages like those of the Vuilleumier concepts, and their operation is similar enough to be analyzed by the same techniques. We selected two Stirling-Stirling concepts as examples:

- A single-cycle free-piston concept, called the duplex Stirling heat pump by its inventor.
- A four-cycle Stirling-Stirling concept similar to the balanced-compounded Vuilleumier heat pump. The inventor called this concept the balanced-compounded Stirling heat pump.

These concepts differ from Vuilleumier heat pumps mainly in how the working fluid moves and how power is transmitted from the engine segment to the refrigerator segment. Although the engine and refrigerator segments are not as fully integrated as in Vuilleumier concepts, they are too closely connected to be classified as split-cycle embodiments. Figure 4.2 shows how the six concepts may be classified according to whether the engine and refrigerator share working fluid and whether they mechanically compress the fluid.

Vuilleumier-cycle heat pump concepts offer full integration of the heat discharged from the engine and the refrigerator. Since this is the heat used for comfort heating, it is very desirable to have both heat flows at the same temperature. The Stirling-Stirling concepts can reject heat from the engine and refrigerator segments at two different temperature levels. Since that has no value for comfort-heating heat pumps, our analysis required that both heat-discharge temperatures be the same. In practice, this is accomplished by properly sizing the heat exchangers, and thermally linking them.

We considered analyzing two other related concepts as examples, but we did not select them. One is the heat-activated regenerative compressor (HARC) concept developed by the Institute of Gas Tech-

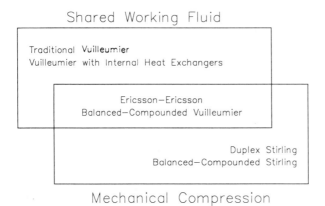

Figure 4.2 Integrated heat pump classification.

nology.[3] The engine segment of this Ericsson-Rankine concept has rotating (radially sweeping) displacers and directly compresses a refrigerant vapor. The other concept was proposed by the late Dr. William Martini, Martini Engineering, Richland, Washington. It couples a free-piston Stirling engine with a Clausius-Rankine refrigeration compressor. Both concepts use check valves to control the direction of refrigerant flow, and both are driven by oscillating members reacting to the engine working-fluid pressure. These concepts have the advantage of using standard hardware for the refrigerator segment and can recover heat from the engine. We did not analyze them because their Clausius-Rankine refrigerators can be analyzed by well-known methods and their engine segments are similar to the engine segments of some of the six concepts selected. The HARC engine segment is similar to the engine segment of the Vuilleumier with internal heat exchangers, and the Stirling-Rankine engine segment is similar to the duplex Stirling engine segment.

Origin and Development of Integrated Heat Pump Concepts

Beginning in the 1950s, several researchers have revisited various heat-activated heat pump concepts, including Vuilleumier machines. The success of Stirling-type concepts in cryogenic applications encouraged the development of similar devices for the higher temperature levels and larger capacities needed for air conditioning. For the past thirty years, the U.S. natural gas industry has led the development of heat-activated heat pumps in order to maintain its dominant position in residential and commercial space heating. Many advances in space-conditioning science and technology have resulted from this research.

The first mechanical and regenerative concepts based on combined cycles considered for air-conditioning applications used thermocompressors, although not of the Vuilleumier type. Several approaches were investigated from the 1950s through the 1970s.[4] Serious development started in the 1960s for Ericsson-cycle machines and in the 1970s for duplex Stirling-cycle machines.

The following sections describe the research and development history of the six heat pump concept embodiments presented in this book. Not all of them have reached the equipment development stage, but they have technical merit, and R&D could reduce them to practice in a reasonable time.

Traditional Vuilleumier heat pump

In 1918, Rudolph Vuilleumier, an engineer in New Rochelle, New York, patented a regenerative refrigeration machine[5] similar to the

Stirling cycle. In doing so, he joined a group of distinguished professionals, including George Brayton (a Boston engineer), James Joule (an English brewer), and others, who, over a short time, proposed significant advances in refrigeration technology that achieved practical use. While others proposed concepts in which a shaft or drive belt physically connected the engine and the refrigerator, Vuilleumier conceived a cooling machine in which the engine and the refrigerator use a common working fluid.

Vuilleumier conceived this deceptively simple combination early in the twentieth century, when the technologies of compression refrigeration, ejector refrigeration, absorption cooling, and air-cycle refrigeration were already mature, and cryogenics was becoming well established. By then, the United States was preeminent in the manufacture of refrigeration machines. Vuilleumier's invention did not successfully compete with these established technologies. We have no evidence that Vuilleumier actually built a heat pump. The application of his concept came later, for the special case of miniature cryogenic refrigerators which needed a simple cycle embodiment more than high efficiency.

The basic Vuilleumier concept is simple, but adequately analyzing the performance of real machines with interacting components is a more formidable task. This difficulty may explain why most thermodynamics and refrigeration textbooks do not describe the cycle and its embodiments. As a result, design engineers are less likely to be familiar with analysis of Vuilleumier cycles and equipment. This unfamiliarity has been one reason why this equipment has not seen wider use. Another reason is that good performance of the equipment relies heavily on near-perfect heat-exchanger performance.

The other five concept embodiments described were conceived and developed exclusively in the United States, but research on the traditional Vuilleumier concept has not been limited to this country. Philips Research Laboratories in Eindhoven, Holland, considered it first for low-temperature refrigeration, and then for heat pumps in the 1970s. Vuilleumier machines have been under evaluation and development since the 1960s in the United States, Germany, Holland, and Japan for heat pumps and air conditioners.

Much of the R&D has focused on multiple-cylinder, single-cycle, constrained-piston, thermal-compression Vuilleumier machines. Philips Laboratories in Briarcliff Manor, New York, and Eindhoven, Holland, focused on Vuilleumier cycles in cryogenic refrigerators[6] to achieve cooling below 100 K ($-280°F$). More recent work at the Danish Invention Center, Technological Institute, in Taastrup, Denmark, has stressed the development of similar machines for space conditioning. The embodiment of this concept, described in Chap. 5, has external heat exchangers and regenerators. Other configurations, including one studied at Briarcliff Manor, use hollow displacers with regenerator material packed inside them. Instead of using heat exchangers, they rely on direct heat transfer through the walls and ends of the cylinder. Some recent developments are described in Chap. 8.

Vuilleumier heat pump with internal heat exchangers

This concept embodiment evolved from a thermal compressor development that began in the 1960s. Research at the Institute of Gas Technology (Chicago) led to the heat-activated regenerative compressor (HARC) embodiment, whose operating cycle was based on thermally compressing and expanding a working fluid.[7,8] The compressor discharged a portion of the working fluid through an automatic valve to act as refrigerant in a conventional Clausius-Rankine refrigerator. This fluid was pulled back into the thermocompressor through an automatic suction valve after it absorbed heat in the evaporator. Because of the valves, the HARC thermocompressor emulates the Ericsson cycle. In a second-generation embodiment, the pressure variations were transmitted through a diaphragm to a free-piston, vapor-compression refrigerator. This machine made innovative use of high-temperature heat pipes.

In a further evolutionary step, the thermocompressor was used as the engine portion of a Vuilleumier cycle, driving a second unit that acted as a refrigerator. This embodiment—the Vuilleumier heat pump with internal heat exchangers—was awarded a U.S. patent in 1984.[9] The Institute of Gas Technology has pursued its further development because of industrial interest. Although the thermocompressor has been tested, performance data are not available for the heat pump system because this research was confidential.

Duplex Stirling heat pump

In the 1960s, Professor William Beale of Sunpower, Inc., developed a Stirling-engine-driven vapor-compression heat pump. Later, he expanded his research to include a Stirling-Stirling machine called the duplex Stirling heat pump. Beale's work is well documented in the literature,[10] and some recent developments are described in Chap. 8.

Dr. Graham Walker had proposed a similar Stirling refrigerator in 1965.[11] The major difference between Beale's and Walker's machines is that Walker used springs for energy storage, and his displacers, instead of being solid, were formed to also serve as regenerators. Although Walker's idea was proposed as a cryogenic refrigerator rather than a heat pump, it represents an advanced concept that can be classified as an integrated heat pump.

Balanced-compounded Stirling heat pump

Dr. Theodor Finkelstein first conceived the idea of multiple-cycle heat pumps whose component cycles are in balanced phasing with each other in 1975. He presented the basic principles of balanced compounding in 1978.[12] These principles underlie both this heat pump and the one described next: the balanced-compounded Vuilleumier heat pump.

Although some work was done with models of the engine segment, we are not aware of any equipment development based on the concept.

Balanced-compounded Vuilleumier heat pump

The basic principles of balanced compounding that Finkelstein described in 1978[12] led to the balanced-compounded Vuilleumier heat pump embodiment that he patented in 1980.[13] The operating characteristics of this heat pump were published in 1980.[2] We are not aware of any equipment development that has been based on this concept.

Ericsson-Ericsson heat pump

The origin of concepts which follow the thermodynamic cycle that was eventually named after John Ericsson can be traced back to his development of a hot-air engine. This engine contained a piston and a regenerative displacer, coupled together by a crank mechanism that moved them 90° out of phase. Unlike in Stirling machines, the fluid flow is controlled by valves.

Many design variations of Ericsson-cycle engines have been proposed.[14] Both piston and displacer configurations are possible, and some of the designs incorporate regenerators into the displacers. Some have achieved commercial application in cryogenic machinery, and others have been developed for use in powering mechanical hearts.[15,16]

Dr. Glendon M. Benson of Energy Research and Generation, Incorporated (ERG) developed a concept known as the Ericsson-Ericsson heat pump as part of a long-term program to develop free-piston engine-driven heat pumps. ERG evaluated Otto-Rankine, Diesel-Rankine, Rankine-Rankine, Stirling-Stirling, Stirling-Rankine, and Ericsson-Ericsson concepts for the gas industry.[1] After doing research for the American Gas Association on Stirling-Rankine concepts, ERG decided to pursue the Ericsson-Ericsson concept. This concept was first described in the literature in 1973.[17] Its embodiment includes gas bearings, and instead of external heat exchangers it uses internal-fin devices that ERG calls thermizers. The thermizers reduce cylinder dead space and create more nearly isothermal conditions in the variable-volume gas spaces. By doing so, they enable a closer approach to the ideal Ericsson cycle than more conventional Ericsson equipment.

The embodiment analyzed in this book is based on a multiple-cylinder, single-cycle, free-piston machine designed by Dr. Benson.[1] ERG has tested at least one machine of this type. Dr. Benson has also designed multiple-cycle versions, and ERG has reported a further advance which integrates linear alternators into the heat pump to form a residential energy system that provides both heat and electric power.[18] Regrettably, progress on this development has not been further reported.

References

1. Benson, G. M., "Free-Piston Heat Pump," paper presented at Institute of Gas Technology Stirling-Cycle Prime Movers Seminar, Rosemont, Ill., June 14–15, 1978. Published in *Seminar Proceedings*, pp. 67–99 (October 1979).
2. Finkelstein, T., "Analysis of a Heat-Activated Stirling Heat Pump," paper No. 809363 presented at the 15th Intersociety Energy Conversion Engineering Conference, Seattle, August 18–22, 1980.
3. Kinast, J. A., and J. Wurm, "Gas-Fired Heat Pump Research and Development—Review and Assessment," final report for Gas Research Institute under contract 5081-242-0541, December 1983, pp. 28–29.
4. Kinast, J. A., and J. Wurm, "Gas-Fired Heat Pump Research and Development—Review and Assessment," final report for Gas Research Institute under contract 5081-242-0541, December 1983, pp. 12–43.
5. Vuilleumier, R., "Method and Apparatus for Inducing Heat Changes," U.S. Patent 1,275,507 (1918).
6. Pitcher, G. K., and F. K. du Pré, "Miniature Vuilleumier-Cycle Refrigerator," *Proceedings of the 1969 Cryogenic Engineering Conference*, pp. 447–451 (1970).
7. Granryd, E. G. V., "Heat-Actuated Regenerative Compressor for Refrigerating Systems," U.S. Patent 3,413,815 (Dec. 3, 1968), (assigned to American Gas Association, Inc.).
8. Granryd, E. G. V., "Heat-Actuated Regenerative Compressor System," U.S. Patent 3,474,641 (Oct. 28, 1969), (assigned to Gas Developments Corporation).
9. Wurm, J., and J. A. Kinast, "Heat Actuated Heat Pumping Apparatus and Process," U.S. Patent 4,455,841 (1984).
10. Penswick, L. B., and I. Urieli, "Duplex Stirling Machines," paper 849045 presented at the 19th Intersociety Energy Conversion Engineering Conference, San Francisco, August 1984.
11. Proposal No. 66-141M, "Miniature Stirling Cycle Cooler for Infra Red Detectors," from IIT Research Institute to U.S. Army E.R.D.L., Far Infrared Laboratory, October 1965.
12. Finkelstein, T., "Balanced Compounding of Stirling Machines," paper 789194 presented at the 13th Intersociety Energy Conversion Engineering Conference, San Diego, August 1978.
13. Finkelstein, T., "Method and Device for Balanced Compounding of Stirling Cycle Machines," U.S. Patent 4,199,945 (1980).
14. Walker, G., *Stirling-Cycle Machines*, London: Oxford University Press, 1973, pp. 61–63.

15. Glassford, A. P. M., "Adiabatic Cycle Analysis for the Valved Thermal Compressor," *Journal of Energy*, 3(5): 306–314 (1979)
16. Walker, G., *Stirling Engines*, London: Oxford University Press, 1980, pp. 17.10–17.17.
17. Benson, G., "Thermal Oscillators," paper 739076 presented at the Intersociety Energy Conversion Engineering Conference, Philadelphia, 1973.
18. Rifkin, W. D., G. M. Benson, and R. J. Vincent, "Applications of Free Piston Stirling Engines," paper 809401 presented at the 15th Intersociety Energy Conversion Engineering Conference, Seattle, August 18–22, 1980.

Chapter

5

Description of Stirling-Stirling and Vuilleumier Heat Pumps

Six embodiments of the Stirling and Vuilleumier heat pumps are described. A schematic representation of each concept is presented, and its operation is explained. The generic advantages and disadvantages of each cycle configuration are discussed, and differences and similarities of these embodiments are detailed.

The six concept embodiments discussed and evaluated, in general order from the simplest to the most complex, are:

1. Traditional Vuilleumier heat pump
2. Vuilleumier heat pump with internal heat exchangers
3. Duplex Stirling heat pump
4. Balanced-compounded Stirling heat pump
5. Balanced-compounded Vuilleumier heat pump
6. Ericsson-Ericsson heat pump

In our analysis, the above cycles operate with three characteristic temperatures. By our definition, Vuilleumier concepts operate over three temperatures. The Stirling-Stirling concepts, which can operate over four temperatures, were constrained to three temperatures by having the engine and refrigerator segments reject heat at the same temperature. The spaces associated with the high temperature, intermediate temperature, and low temperature are typically referred to as the hot, warm, and cold spaces, respectively. Heat transfer between the external environment and the working fluid usually occurs at the heat exchangers (unless a special configuration is used). Working-fluid expansion and compression is nearly adiabatic, changing temperature as the pressure varies.

To describe how the six concepts work, a specific mechanical embodiment is portrayed in each of the following sections. These embodiments do not represent optimal equipment designs. They were selected for clarity of explanation of how a mechanical embodiment based on these cycles works, so that a general comparative analysis of these integrated heat pumps can be made. The schematics typically depict the engine and refrigerator volumes as approximately equal. These design volumes can vary, with their ratio dependent on the operating temperatures of the machine. The design engineer has a wide scope for creatively selecting the operating regime and specific components, and assembling them into an integrated heat pump.

Traditional Vuilleumier Heat Pump

Description

Figure 5.1 shows a schematic of the traditional Vuilleumier configuration. This configuration has reciprocating displacers and external heat exchangers and regenerators, which is the classical form of Vuilleumier machines. It is the form originally proposed by Vuilleumier and has been the basic configuration used by many others in their research. It has mechanically constrained displacers in

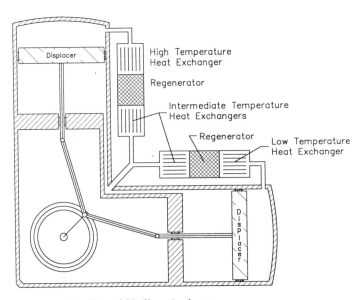

Figure 5.1 Traditional Vuilleumier heat pump.

cylinders. Although it is usually called a piston configuration, the moving elements have only a small pressure difference across them and are more properly called displacers.

These constant-volume thermal-compression machines produce the pressure changes required to drive the cycle by varying the bulk average temperature of the working fluid. This is done by the displacers, which vary the volume of the spaces devoted to each temperature. The total working volume, however, does not change. In contrast, mechanical-compression machines change the pressure of the working fluid by varying both the average temperature and the total volume occupied by the working fluid, with the volume variation controlled by the movement of pistons.

The maximum pressure ratio in thermal-compression machines depends on the operating temperatures in the machine. For the same temperature ratios, thermal-compression machines have lower pressure ratios than mechanical-compression machines operating between the same temperatures. Although the lower pressure ratios produce lower specific capacities, this is not necessarily detrimental. Because small-capacity thermal-compression systems are larger than their mechanical-compression counterparts, they may be easier to build and easier to integrate with the required heat exchangers and regenerators.

Operation

The traditional Vuilleumier machine in Fig. 5.1 has two cylinders, with a double-acting displacer in each cylinder. The displacers are linked so that one lags the other in phase by 90°, which is about optimum, according to Pitcher and du Pré.[1] This phase lag is commonly achieved by placing the cylinders in a 90°-V arrangement and driving the displacers from a common crankshaft. Figure 5.2 shows the motion of the displacers as a progression through four states of the cycle. The first state follows the fourth, closing the cycle.

The two cylinders are identified by their functions. The cylinder associated with the high and intermediate temperatures is referred to as the *engine cylinder*, and the cylinder associated with the intermediate and low temperatures is referred to as the *refrigerator cylinder*. The engine cylinder (on the left in Fig. 5.2) accepts heat at the high temperature, rejects heat at the intermediate temperature, and produces the pressure variation which drives the refrigeration process. The refrigerator cylinder accepts heat at the low temperature (producing a cooling effect), rejects heat at the intermediate temperature, and accepts the pressure variation from the engine segment. The heating effect is obtained from the heat rejected by both segments at the intermediate temperature.

The pressure variation between the engine and refrigerator segments is equalized by a working-fluid connection between them. The

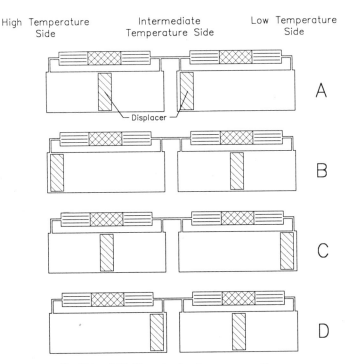

Figure 5.2 Traditional Vuilleumier heat pump operation.

working fluid actually changes segments during operation, driven by the pressure variation between segments. The varying amount of working fluid in the engine and refrigerator segments is the means for transferring power between them.

The pressure variation is caused by varying the average temperature of the entire working fluid without any change in working volume. In state A, the displacer in the engine segment is centered, dividing its volumes evenly between the high and intermediate temperatures. The displacer in the refrigerator segment gives most of the volume to the cold space. The mean gas temperature (weighted average based on the gas distribution among the three temperature levels) is relatively low compared with the rest of the cycle. As a result of the low average temperature, the gas pressure is also relatively low.

In step A to B, the hot space volume is decreased by the displacers, which push the gas toward the warm space. As it moves through the regenerator, the gas leaves heat in the regenerator matrix and enters the warm space near the intermediate temperature. The cold gas is also pushed to its warm space, absorbing heat from the regenerator while passing from the low- to the intermediate-temperature level. The average gas temperature drops somewhat from that of state A, because the intermediate temperature is much closer to the low temperature than the high temperature.

In state C, the hot-space volume has increased and the cold-space volume has decreased. With more of the gas occupying the hot space, the average gas temperature is higher, causing an increase in pressure. Since the overall volume remains the same, the pressure increase causes the gas in each volume to increase in temperature (the adiabatic compression effect). As the temperature of the gas rises, it rejects heat to the outside. Since most of the gas is in the warm space, the heat rejection is primarily from the heat exchangers corresponding to that space.

In step C to D, the hot volume continues to increase, and the cold volume begins increasing. As these volumes increase, more gas is shuttled to the extreme-temperature spaces. As the warm gas travels to the hot space, it absorbs the heat that it left in the engine regenerator when going from the hot to the warm space. It also leaves heat in the refrigerator regenerator when it passes from the warm to the cold space. This heat is later absorbed by the gas when it returns through the regenerator.

In step D to A, the hot space decreases in volume and the cold-space volume increases. The net effect is to decrease the average gas temperature and pressure. With the decrease in pressure, the temperature of each space drops, causing the gas to absorb heat from the outside into the hot and cold spaces, where most of the gas then resides.

Throughout the cycle, the overall volume does not change. The pressure variation during the cycle is caused by the changes in the average temperature of the working fluid as its distribution among the three temperature spaces changes.

Because no volume change occurs, no net external work is produced, although work is transferred between the engine and refrigerator segments. A small amount of external work must be supplied to move the displacers. This work is not required to compress or expand the working fluid, but merely to move it from one space to another. The amount of work necessary can be very small, because it is needed only to overcome fluid friction (flow losses) and mechanical friction.

Vuilleumier Heat Pump with Internal Heat Exchangers

Description

The Vuilleumier configuration shown in Fig. 5.3 is functionally the same as the above embodiment, but it is mechanically different. It has two cylinders with a displacer in each, again resulting in four volumes associated with three temperature levels. The displacers move radially rather than axially, and the heat exchangers and regenerators

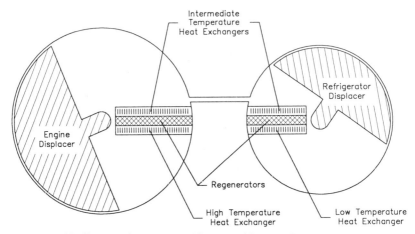

Figure 5.3 Vuilleumier heat pump with internal heat exchangers.

are inside the cylinders. The use of radial displacers is feasible because the displacers are merely moving the working fluid, not compressing and expanding it. Connecting passages, along with their corresponding dead spaces, are reduced or eliminated. These features may permit better working-fluid flow through the heat exchangers.

This embodiment, like the previous one, is a constant-volume machine. The pressure changes required to produce the cycle's effect are achieved by varying the bulk average temperature of the working fluid, with the displacers varying the volumes devoted to each temperature. Like the previous concept, this Vuilleumier embodiment has lower pressure ratios and specific capacities than mechanical-compression machines.

Operation

The steps in this concept's operation are fundamentally the same as for the traditional Vuilleumier. The cylinder associated with the high and intermediate temperatures is referred to as the *engine cylinder*, and the cylinder associated with the intermediate and low temperatures is referred to as the *refrigerator cylinder*. Figure 5.4 shows the displacer motion through one cycle, equivalent to the states in Fig. 5.2. To follow its operation, refer to the text in the preceding section describing the operation of the traditional Vuilleumier.

Duplex Stirling Heat Pump

Description

The duplex Stirling is inherently a single-cycle, single-cylinder, free-piston machine. It can be described as a free-piston arrangement of a

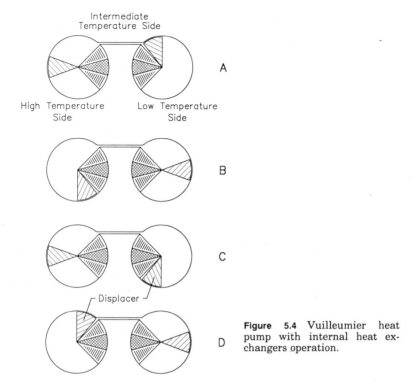

Figure 5.4 Vuilleumier heat pump with internal heat exchangers operation.

displacer-piston Stirling engine and a displacer-piston Stirling refrigerator. In this embodiment, the double-acting piston is shared between the two segments. There are no major mechanical links guiding the piston or displacers. The pistons or displacers are controlled by springs (gas or mechanical) and gas forces.

The piston transfers work between the engine and refrigerator cycles. Energy is stored in the form of a moving mass (the piston separating the two segments) for use during the portion of the cycle when the engine is absorbing work. This stored energy compensates for the different instantaneous requirements of the power and refrigeration cycles. The energy-storage piston is a major factor in determining the size of the duplex Stirling heat pump. The maximum size of the heat pump is limited by the largest mass that can be controlled in a free-piston configuration.

Figure 5.5 shows the components of a duplex Stirling heat pump. One engine and one refrigerator are linked together. More engine and refrigerator units can be added in parallel if more capacity is needed, but they would operate independently.

The duplex Stirling is a mechanical-compression machine. The piston varies the volume occupied by the working fluids of the engine and

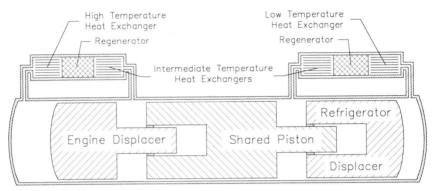

Figure 5.5 Duplex Stirling heat pump.

refrigerator. It has a higher pressure ratio and higher specific capacity than comparable thermal-compression machines.

Engine operation

In the duplex Stirling schematic shown in Fig. 5.5, the left half is the engine segment, and the right half is the refrigerator segment. The ends of the cylinders are associated with the extreme temperatures, while the inner portions (adjacent to the piston) are associated with the intermediate temperature. The displacers are lightweight elements, with no designed pressure difference across them. The working fluid can flow freely from one space to the other through the heat exchangers and regenerators, and any pressure difference would be merely a result of the fluid flow friction in the heat exchangers. The piston, on the other hand, must maintain the full pressure difference between the engine and refrigerator segments. Ideally, no working fluid flows between the segments. In practice, the two segments will have the same mean operating pressure, because gas will leak from one to the other during part of the cycle until the leakage is offset during the remaining part by leakage in the opposite direction.

Figure 5.6 shows the operation of the engine segment. In the diagram for state A, the displacer is at its leftmost position and the piston is at its middle position. In step A to B, the piston moves to the left, and the displacer moves to the right. This mechanically compresses the engine's working-fluid space and pushes the gas through the heat exchangers, where it rejects heat at the intermediate temperature. An important part of this step is the shift of the displacer, which transfers the working fluid from the warm space between the displacer and the piston to the hot space to the left of the displacer. This heating of the gas causes additional compression. As this occurs, the working fluid is

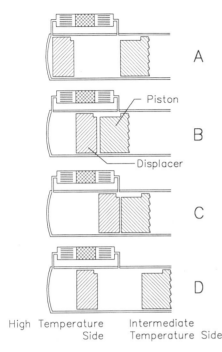

Figure 5.6 Duplex Stirling engine operation.

passing through the regenerator, absorbing heat, so that it approaches the high temperature. There is a 90° phase lag in the displacer and piston motion, similar to the phase lag of the two displacers in the traditional Vuilleumier. The displacer reaches its leftmost position in state A, while the piston reaches its leftmost position in state B.

In step B to C, both piston and displacer travel to the right, expanding the space devoted to the engine working fluid. As this occurs, the gas cools by adiabatic expansion, so that it can absorb heat at the high-temperature level during step C to D. Work is produced by the gas acting on the piston, moving it to the right against the pressure of the refrigerator working fluid on the other side.

In step C to D, the displacer reverses its motion and travels to the left, towards the hot end. This moves the working fluid from the hot space through the regenerator to the warm space. As the fluid traverses the regenerator, it leaves the heat that it will later absorb during step A to B.

To complete the cycle, in step D to A, the displacer moves to the left side of the cylinder, pushing all the working fluid from the hot space to the warm space. The piston begins compressing the engine working fluid, reducing its total volume. The fluid rejects the heat of compression at the intermediate temperature.

The engine produces more work during the expansion steps B to C to D than it absorbs during the compression steps D to A to B, because its mean temperature is higher during the expansion. Its primary heat absorption occurs while most of the gas is in the hot space, and it rejects heat while most of the gas is in the warm space. The difference in the amount of heat transferred corresponds to the amount of work produced.

Refrigerator operation

The operation of the refrigerator segment of the duplex Stirling is shown in Fig. 5.7. These state diagrams correspond to the diagrams for the engine's operation. In state A, the piston is at its midpoint, moving to the left, and the displacer is in the farthest left position. In step A to B, the displacer moves toward the cold end, pushing the working fluid from the cold space through the heat exchangers and the regenerator to the warm space. As the fluid traverses the regen-

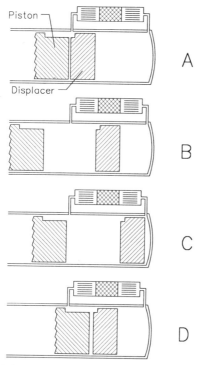

Figure 5.7 Duplex Stirling refrigerator operation.

erator, it absorbs heat and enters the warm space near the intermediate temperature.

In step B to C, the piston begins compressing the working fluid, whose transfer to the warm space is completed by the displacer.

The heat of compression is rejected through the warm heat exchanger as the gas is displaced through it in step C to D. In this step, the piston completes its compression stroke and the displacer moves toward the left, pushing the working fluid from the warm space to the cold space.

Completing the cycle, in step D to A, the piston begins moving to the left, expanding the volume of the refrigerant segment. As the gas expands, it cools and absorbs heat while it passes through the cold heat exchanger.

The refrigerator produces work during its expansion steps D to A to B, but it absorbs still more work during its compression steps B to C to D. Heat is absorbed by the gas while it is primarily at the low temperature, and is rejected while it is primarily at the intermediate temperature. While the refrigerator absorbs work during B to C to D, the engine produces work during these steps, and the work absorbed by the engine during its D to A to B compression is produced by the refrigerator expansion. Achieving this balance in practice is a major design consideration for this concept. Some of the mechanical energy must be briefly stored in the inertia of the piston and in the gas springs described below.

Gas spring operation

There is an inherent phase difference between the piston and the displacers. The piston completes a compression or expansion over two steps in the above diagrams, while the displacers' function has a phase lag of 90°. The duplex Stirling achieves this phase difference by using a heavy piston and lightweight displacers. The lightweight displacers respond quickly to gas pressures acting on them, while the piston responds slowly because of its inertia. Referring back to Fig. 5.5, the gas pressures acting on the displacers are the working-fluid pressures on their faces and also the pressures in the smaller spaces where the piston and displacers fit together. With adequate sealing, the gas in these spaces does not interchange with the working fluid. This provides a gas spring or bounce space. Although, as pictured, the engine displacer has a rod extending into the piston, and the piston has a rod extending into the refrigerator displacer, the rods could extend in either direction. Other embodiments are also possible, including crank-constrained pistons and displacers.

The key to this embodiment's operation is that the gas springs provide an opposing force that lags in phase with the operating pressures in the engine and refrigerator. For example, as the engine working-fluid temperature rises, the temperature in the gas spring does not rise until the engine displacer begins to move toward the piston. Thus, there is a net force on the displacer in the direction of the piston until the gas spring is compressed sufficiently that the pressure in the spring equals the hot-space pressure. When the engine's gas spring pressure becomes large enough, the piston will begin to move toward the refrigerator displacer. This piston motion compresses the refrigerator's gas spring, producing a net force on the refrigerator displacer that causes it to move. The pressure in the cylinder oscillates above and below the bounce-space pressure, causing the displacer and piston to move together or apart, respectively. Gas springs are also used in an equivalent manner in the Ericsson-Ericsson heat pump.

Balanced-Compounded Stirling Heat Pump

Description

This embodiment is a four-cycle, eight-cylinder, free-piston machine. To produce the correct phase relationship between pistons, four Stirling engines and four Stirling refrigerators are connected in a Siemens arrangement. The Siemens configuration was first devised by the British scientist and engineer Sir William Siemens, and later reinvented by F. L. van Weenan of Philips Research Laboratories. It consists of multiple cylinders and double-acting pistons. The working fluid moves back and forth between adjacent cylinders, passing through two heat exchangers and a regenerator. The most common Siemens engine configuration uses four cylinders which follow four identical Stirling cycles, each operating 90° out of phase with its neighbors.

The balanced-compounded Stirling is a mechanical-compression machine. As the piston moves, it changes the volume occupied by the working fluid, mechanically expanding and compressing the gas. Because the balanced-compounded Stirling uses mechanical compression in addition to varying the average temperature of the working fluid, it has a higher pressure ratio and higher specific capacity than thermal-compression machines. This translates into smaller machines for a given capacity, important for producing high-capacity equipment.

General operation

The balanced-compounded Stirling is shown in Fig. 5.8. It has eight cylinders, each with a double-acting piston. Four cylinders are devoted to engine cycles; the other four are devoted to refrigeration cy-

Description of Stirling-Stirling and Vuilleumier Heat Pumps

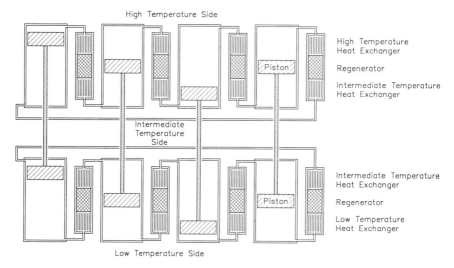

Figure 5.8 Balanced-compounded Stirling heat pump.

cles. One side of each engine piston acts on a hot space, while the other side acts on a warm space. One side of each refrigerator piston acts on a warm space, while the other side acts on a cold space. Each engine piston and cylinder is mechanically linked to a refrigerator piston and cylinder, so that the warm spaces are close to each other. Increasing the volume of an engine warm space decreases the volume of a refrigerator warm space. There are flow connections between the warm space of each cylinder and the hot space (for an engine) or cold space (for a refrigerator) of an adjacent cylinder. The motion of the pistons is arranged (either by mechanical linkages or by free-piston dynamics) so that the changing volumes of the hot and cold spaces are out of phase with the volume changes in the warm spaces. The hot and cold spaces reach their minimum volume one-quarter cycle before the warm spaces.

As noted above, the working-fluid path of each segment, engine and refrigerator, follows the four-cylinder Siemens configuration. One way to couple the two segments is with a shaft that transfers power between the engine and refrigerator segments. However, the embodiment analyzed in this book directly couples each piston of the engine to a corresponding piston in the refrigerator segment.

Engine operation

Figure 5.9 shows the operation of one engine cycle. The other three pairwise combinations of engine cylinders follow the same cycle, separated by a 90° phase angle. In each state diagram, the left cylinder is

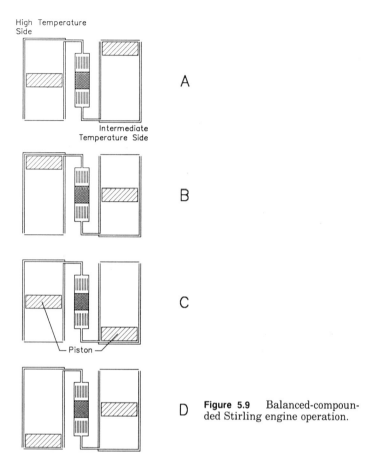

Figure 5.9 Balanced-compounded Stirling engine operation.

the hot space of the engine cycle, and the right cylinder is the warm space. (Although not shown, the left cylinder also holds the warm space for the earlier cycle, and the right cylinder holds the hot space for the later cycle.)

State A shows the high-temperature piston at its middle position and the intermediate-temperature piston at its highest position, creating the maximum volume for the warm space. In step A to B, the high-temperature piston moves the gas from the hot space to the warm space. At the same time, the intermediate-temperature piston begins compressing the working fluid in its space. The heat of this compression is rejected from the gas through the warm heat exchanger during step B to C. During step A to B, the working volume drops from 1.5 to 0.5 cylinder volume.

In step B to C, the intermediate-temperature piston continues reducing the warm-space volume, while the high-temperature piston moves downward, increasing the volume of the hot space. This shifts the working fluid from the warm space to the hot space. The moving fluid absorbs heat from the regenerator, so that when it enters the hot space, it is at the high temperature. The total space devoted to the working fluid remains 0.5 cylinder volume.

In step C to D, the intermediate-temperature piston has completed its downward travel and begins to move upward, increasing the warm volume. The high-temperature piston continues moving downward, and the working space expands from 0.5 to 1.5 cylinder volumes. This expansion cools the gas, and heat will be absorbed by the working fluid as it passes through the hot heat exchanger in step D to A.

Step D to A completes the cycle. The high-temperature piston moves from its lowest position back to the midpoint of its travel, and the intermediate-temperature piston moves from its midpoint to its lowest position. The overall volume remains 1.5 cylinder volumes, but the working fluid has moved from the hot space through the heat exchangers and the regenerator to the warm space. The regenerator absorbs heat from the gas, which enters the warm space near the intermediate temperature.

The engine produces work during the expansion step, C to D, while the working fluid is relatively hotter, and it consumes a smaller amount of work during the compression step, A to B.

Refrigerator operation

Figure 5.10 shows the refrigerator-cycle steps. Like the engine, the refrigerator segment has four identical cycles which are duplicate, separated in phase by 90°. The piston locations in each state diagram correspond to the piston locations in the engine-cycle diagram. The left cylinder is associated with the cold space, and the right cylinder is associated with the warm space.

In state A, the low-temperature piston is at its midpoint and the intermediate-temperature piston is at its highest position. All the working fluid is in the cold space. In step A to B, the low-temperature piston increases the volume of the cold space, and the intermediate-temperature piston moves to its midpoint, increasing the volume of the warm space. The total working volume increases from 0.5 to 1.5 cylinder volumes. The expansion cools the working fluid, which can then absorb heat in the cold heat exchanger during step B to C.

In step B to C, the working fluid is pushed from the cold space to the warm space as the pistons move. The gas moves through the regenerator, where it absorbs heat, and enters the right cylinder near the in-

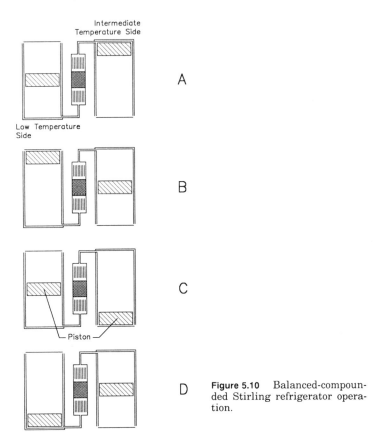

Figure 5.10 Balanced-compounded Stirling refrigerator operation.

termediate temperature. The total working volume is still 1.5 cylinder volumes.

In step C to D, the low-temperature piston completes its reduction of the cold-space volume, pushing all the working fluid to the warm-space volume. At the same time, the intermediate-temperature piston compresses the working fluid, reducing the total volume devoted to the gas from 1.5 to 0.5 cylinder volume. The compressed working fluid rejects the heat of compression at the warm heat exchanger during step D to A.

Step D to A completes the cycle. The working fluid is pushed from the warm space back to the cold space. The regenerator absorbs heat from the fluid, so it enters the cold space near the low temperature. The heat that is left in the regenerator in this step is reabsorbed by the working fluid during step B to C.

The refrigerator consumes work during the compression step, C to D, and generates a smaller amount of work during the cooler expansion step, A to B. The net work generated by each engine cycle is equal to the work absorbed by each refrigerator cycle. Furthermore, an additional feature, identified as especially desirable by the inventor, Dr. Finkelstein,[2] is that the work-consuming step of each cycle coincides with the work-producing step of the cycle that is linked to it by the piston link. As a result, very little work needs to be stored in the system by a flywheel mechanism.

Balanced-Compounded Vuilleumier Heat Pump

Description

This heat pump is a six-cylinder, four-cycle, free-piston machine. Each cycle has two expansion spaces, one for the engine segment and one for the refrigerator segment, and one shared compression space. It has three characteristic temperature levels, as do the other Vuilleumier embodiments. Finkelstein's design[3] is related to the Stirling-Stirling concepts because, unlike the other Vuilleumier embodiments, it uses mechanical compression. It differs from the balanced-compounded Stirling in that the working fluid in the Vuilleumier can be in a space associated with any of the three temperature levels, whereas the Stirling keeps the working fluid in the engine segment (hot and warm spaces) physically separate from that in the refrigerator segment (warm and cold spaces).

To produce the correct phase relationship between pistons, the four Vuilleumier cycles are connected in a similar manner to the balanced-compounded Stirling. Double-acting pistons are used, with the higher-temperature side of one working space sharing a piston with the lower-temperature side of the adjacent working space. Unlike in the balanced-compounded Stirling, the power generated by the engine segments is transferred by the common working fluid directly acting on the working fluid of the refrigerator segments. This reduces the number of pistons by two and the number of heat exchangers by four. Figure 5.11 is a schematic diagram for the balanced-compounded Vuilleumier. In this embodiment, the two free-piston assemblies are in sustained thermally activated oscillation. The three bands of heat exchangers and their related spaces are, in order from the outside to the inside, hot, warm, and cold. The crossed-cylinder arrangement in this figure clarifies the flow and piston connections and the basic symmetry of this configuration. In a real system, the cylinders would be side by side or in some other convenient position.

According to Dr. Finkelstein, a balanced-compounded Vuilleumier can be configured in either free-piston or constrained-piston form. Our

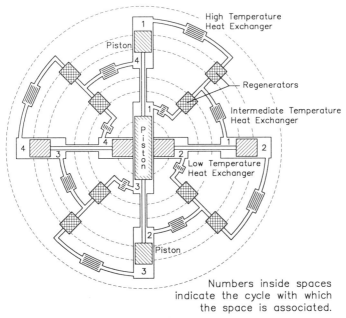

Figure 5.11 Balanced-compounded Vuilleumier heat pump.

analysis of the balanced-compounded Vuilleumier presumes sinusoidal motion of the pistons and therefore applies to either form. A more complete study of the motion of the elements would be needed for actual hardware design.

Unlike the other Vuilleumier embodiments, the balanced-compounded Vuilleumier is a mechanical-compression machine. It is an embodiment of a Vuilleumier concept because the same element of working fluid can be at any of three different temperatures, whereas in a Stirling, a working-fluid element can be at only one of two temperatures. Like the balanced-compounded Stirling, its pistons vary the volume occupied by the working fluids. It has a higher pressure ratio and higher specific capacity than the thermal-compression Vuilleumier machines. This higher specific capacity enables smaller machines to be produced for a given refrigeration capacity, which is important for achieving reasonable equipment sizes.

This concept is a relatively straightforward combination of four Vuilleumier-cycle heat pumps. The machine theoretically requires no drive mechanisms or valves. The only moving parts are the two piston assemblies, which, in a free-piston configuration, oscillate at their natural frequency, driven by pressure forces exerted by the working fluid. Alternatively, the piston assemblies can be constrained by a crank or equivalent mechanism. Constrained pistons, controlled by the crank, would follow sinusoidal motion more closely and not change their stroke when operating conditions change. The basic operation of the free-piston and constrained-piston versions is the same. Their thermodynamic differences become more distinct at off-design operating conditions.

The free-piston form contains two reciprocating piston assemblies (cross-hatched in the figure), each consisting of three double-acting pistons connected by piston rods. There are no external linkages to these free-piston assemblies, nor are there any internal drive mechanisms. Each piston assembly is mounted in three in-line cylinders, with close-fitting interconnecting passages to accommodate the piston rods. Each outer cylinder has a high-temperature end and an intermediate-temperature end. Both ends of the inner cylinders are low-temperature ends. The piston assemblies move 90° out of phase with each other. When one assembly is at either end, the other is positioned at the midpoint of its travel. (To allow labeling of the spaces in Fig. 5.11, this positioning is not shown.)

As noted above, the heat pump consists of four identical cycles, designated 1, 2, 3, and 4 in the figure. The four cycles contain identical charges of the working fluid. They go through identical thermodynamic cycles, with a 90° phase difference between adjacent quadrants.

Each Vuilleumier cycle consists of three variable-volume gas spaces: a hot space, a cold space on the same half of one piston assembly, and a warm space in contact with the adjacent piston assembly. Working fluid flows among these three spaces through three heat exchangers, one associated with each temperature level, and through two regenerators.

Since the four cycles are identical and their working-fluid charges are separate (except for minor leakage past the pistons), this description of the operation focuses on one cycle. The other three are equivalent except for their phase difference. The cycle contains a hot space where heat is supplied to the cycle, a cold space where heat is absorbed by the heat pump, and a warm space where heat is rejected by the heat pump.

Each Vuilleumier cycle can be visualized in two segments, an engine and a refrigerator. Figures 5.12 and 5.13 show each segment's operation through one complete cycle.

Engine operation

The engine segment contains a hot expansion space and a warm compression space. Their volume variations at quarter-cycle intervals for steady-state operation are illustrated in Fig. 5.12.

In state A, the warm compression-space piston is at its mid-stroke

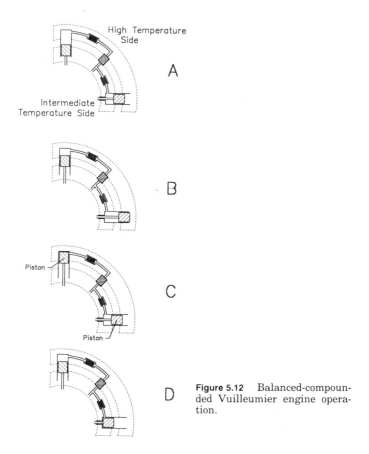

Figure 5.12 Balanced-compounded Vuilleumier engine operation.

position, and the hot expansion-space piston is at its lowest position. In step A to B, the hot expansion space diminishes in volume, and the warm compression-space volume correspondingly increases. This pushes the hot working fluid through the engine regenerator, storing heat there, and into the warm space near the intermediate temperature.

In step B to C, the rest of the working fluid is moved from the hot space to the warm space, and in state C the working fluid begins to be compressed as the warm compression-space piston does work on the gas. This compression causes heat to be rejected from the working fluid through the warm heat exchangers during step C to D.

In step C to D, the motion of the pistons forces the working fluid from the warm space back to the hot space. In passing through the regenerator, the working fluid absorbs heat previously stored there in step A to B, heating the fluid back to near the high temperature.

Step D to A completes the cycle. The hot working fluid expands in

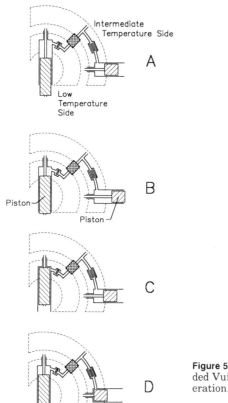

Figure 5.13 Balanced-compounded Vuilleumier refrigerator operation.

the hot cylinder, doing work on the high-temperature expansion piston. This expansion cools the gas, so it will absorb heat in the hot heat exchanger during step A to B. This hot expansion produces more work than is consumed in the warm compression of step B to C. The net work produced by the engine drives the refrigerator.

Refrigerator operation

The refrigerator segment contains a cold expansion space and a warm compression space. Fig. 5.13 illustrates their volume variation at quarter-cycle intervals (at states that correspond to the above engine states).

In step A to B, the movement of the pistons pushes the gas from the cold expansion space to the warm compression space. While passing through the refrigerator regenerator, the gas absorbs the heat which was stored there during step C to D.

In step B to C, the working fluid is pushed from the cold expansion space and compressed. This requires work on the gas, and, as the gas is compressed, its temperature rises, so it will reject heat through the warm heat exchanger during step C to D.

In step C to D, the compressed gas is transferred from the warm compression space to the cold expansion space through the refrigerator regenerator, which absorbs heat and cools the gas to the temperature of the cold space.

In step D to A, the gas in the cold space expands, cooling and delivering work. The expanded gas absorbs heat through the cold heat exchanger during step A to B. This cold expansion produces less work than is consumed by the warm compression of step B to C. In steady operation, the net power produced by the engine balances the net power consumed by the refrigerator.

Like all Vuilleumier machines, the balanced-compounded Vuilleumier attempts to distribute the working fluid throughout the three temperature levels so that the pressure is uniform at any given instant. Cycles in phase with each other have the same pressure levels at any given time. However, coupling multiple cycles so that they are out of phase with each other produces pressure differences, which drive the pistons. This makes the balanced-compounded Vuilleumier self-actuating, unlike the single-cycle Vuilleumier embodiments described above, which require an outside force to move the displacers.

Ericsson-Ericsson Heat Pump

Description

Stirling engines are traditionally defined as regenerative engines where the flow is controlled by volume changes. In comparison, Ericsson engines are usually defined as regenerative engines where the flow is controlled by valves. These heat pumps are substantially different from all of the above heat pumps because of the interruption of flow by the valves or equivalent components.

This embodiment is a three-cylinder, single-cycle, free-piston machine, as shown in Fig. 5.14.[4] It does not strictly adhere to the Ericsson requirement of valves controlling the flow. However, the inventor and developer of the concept, Dr. Glendon Benson, chose the Ericsson-Ericsson designation to differentiate it from the Stirling-Stirling machines he was studying. The Ericsson cycle, which has isothermal compression with constant pressure regeneration, was selected to ideally represent the real machine because it closely approximates the working-fluid cycle.

For thermodynamic comparison, this embodiment is more accurately described as a mechanical-compression Vuilleumier, similar to

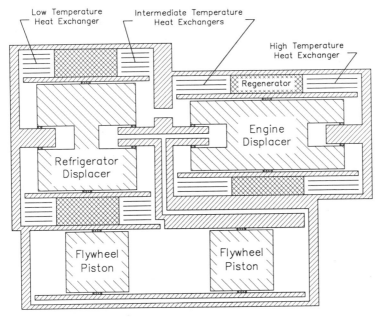

Figure 5.14 Ericsson-Ericsson heat pump.

one of the four cycles of the balanced-compounded Vuilleumier heat pump. The various spaces are interconnected, so the working fluid can be at any of three temperatures, consistent with the definition of a Vuilleumier machine. The pistons vary the overall operating volume of the working fluid, as in other mechanical-compression heat pumps. This embodiment can have either a free- or constrained-piston configuration. It also delivers the relatively higher pressure ratio and specific capacity of mechanical-compression machines.

The Ericsson-Ericsson heat pump has an engine displacer, a refrigerator displacer, and two flywheel pistons, as shown in Fig. 5.14. The opposed orientation of the flywheel pistons and displacers provides dynamic balancing. According to the inventor, the engine and refrigerator segments act on the common working fluid, which, in the limits of discontinuous motion of the moving elements and isothermal variable-volume chambers, provides a classic combination of an Ericsson engine and an Ericsson refrigerator cycle.

Operation

Figure 5.15 shows the movement of the pistons and displacers. In state A, the displacers are at their outermost positions, so the working fluid is primarily in the warm spaces. The flywheel pistons are also at

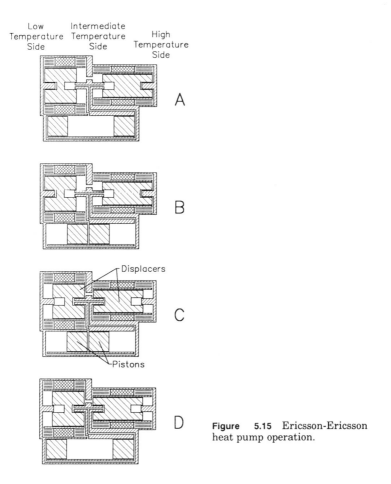

Figure 5.15 Ericsson-Ericsson heat pump operation.

their outermost positions, opening the space between them to the maximum and allowing the working fluid to fill this space. In step A to B, the pistons come together, compressing the working fluid and raising its temperature. The fluid rejects the heat of compression through the warm heat exchangers during step B to C. During this step, work is done by the pistons on the working fluid in compressing it.

In step B to C, the displacers push the working fluid from the warm spaces through the regenerators into the hot and cold spaces. The fluid in the refrigerator segment loses heat to the refrigerator regenerator and enters the cold space near the low temperature. The fluid in the engine segment absorbs heat from its regenerator.

In step C to D, the working fluid expands because its average temperature is higher, and it pushes the flywheel pistons apart. This expansion cools the working fluid, which can then absorb heat in the hot and cold heat exchangers during step D to A.

Step D to A completes the cycle. The displacers move back to their locations in state A, pushing the working fluid from the cold and hot spaces back to the warm spaces. As it passes through the regenerator, the working fluid in the refrigerator segment absorbs the heat left there in step B to C, and the working fluid in the engine segment leaves heat that it will absorb from the engine regenerator during step B to C.

Gas spring operation

For the system to operate without external work input, the work that is produced during the expansion step C to D must equal the work consumed during the compression step A to B. This free-piston Ericsson-Ericsson heat pump has no cranks or other mechanical regulating mechanisms. The moving elements interact by means of the working fluid. The relatively lightweight displacers quickly respond to the pressure differences. As indicated in Fig. 5.14, the hot and cold ends of the displacers have sealed chambers which act as gas springs. At the warm ends, chambers equivalent to the gas springs are in communication with each other and with the spaces on the outside ends of the flywheel pistons. The instantaneous pressure in this connected space is 180° out of phase with the pressure in the warm space. When the pistons come together in step A to B, the expansion of their outer spaces causes the pressure in the connected spaces to decrease. The pressure difference between the gas spring chambers and the connected space drives the displacers toward each other in step B to C.

In step C to D, when the pistons move apart, they compress the gas in the connected space and drive the displacers back toward the gas springs. This causes the displacer movement of step D to A.

Dr. Benson called the pistons *flywheel pistons* because they store mechanical energy, like the flywheel of an internal combustion engine. As the working fluid between the pistons expands, it moves the pistons apart, storing energy in their motion. When the pressures balance across the pistons, they begin slowing down. By the time they have stopped and reversed direction, they have compressed the gas in their outer spaces (the connected space). This pressure then forces the pistons back together to compress the working fluid between them. He also refers to them as *phasors* because they provide the necessary phase difference between piston and displacer motion, similar to the shared piston of the duplex Stirling.

Summary

Table 5.1 summarizes the configuration and basic operating parameters for each of the six concepts. It describes the basic configuration of each concept and its principal characteristics. Specifically, *cycle* refers to the basic thermodynamic cycle upon which the concept is based. *Configuration* describes the arrangement of the components. *Heat ex-*

TABLE 5.1 Characteristics of Vuilleumier and Stirling Heat Pumps

	Traditional Vuilleumier
Cycle	Vuilleumier
Configuration	Axial, displacer in cylinder
Heat exchange	External
Work transfer	Through fluid
Pressure ratio & specific capacity	Low
External drive	Required
Mechanical energy storage	None required

	Vuilleumier with internal heat exchangers
Cycle	Vuilleumier
Configuration	Radial, displacer (radial sweeping) in cylinder, internal heat exchangers
Heat exchange	Internal
Work transfer	Through fluid
Pressure ratio & specific capacity	Low
External drive	Required
Mechanical energy storage	None required

	Duplex Stirling
Cycle	Stirling engine, Stirling refrigerator
Configuration	Axial, displacer/piston/displacer in cylinder
Heat exchange	External or internal
Work transfer	Through power piston
Pressure ratio & specific capacity	High
External drive	Not required
Mechanical energy storage	In power piston

	Balanced-compounded Stirling
Cycle	Four Stirling engines, four Stirling refrigerators
Configuration	Axial, multiple pistons in cylinder
Heat exchange	External
Work transfer	Through shafts connecting engine segments to refrigerator segments
Pressure ratio & specific capacity	High
External drive	Not required
Mechanical energy storage	In pistons

TABLE 5.1 Characteristics of Vuilleumier and Stirling Heat Pumps *(Continued)*

Balanced-compounded Vuilleumier

Cycle	Four Vuilleumiers, augmented with mechanical compression
Configuration	Axial, multiple pistons in cylinder
Heat exchange	External
Work transfer	Through working-fluid exchange between engine and refrigerator segments, and through shafts between cycles
Pressure ratio & specific capacity	High
External drive	Not required
Mechanical energy storage	In pistons

Ericsson-Ericsson

Cycle	Ericsson engine, Ericsson refrigerator (more accurately represented as mechanical-compression Vuilleumier)
Configuration	Axial, two displacers in one cylinder, two pistons in second cylinder
Heat exchange	External
Work transfer	Through working-fluid exchange between engine and refrigerator segments
Pressure ratio & specific capacity	High
External drive	Not required
Mechanical energy storage	In pistons

change refers to the placement of the regenerators and heat exchangers, with "internal" indicating that these components are housed within the basic shell of the concept and "external" indicating that they are outside the shell and must link with the expansion and compression spaces through connecting passages. *Work transfer* describes the route by which work is exchanged between the engine segment and the refrigerator segment. The Vuilleumier and Ericsson-Ericsson concepts all rely on the transfer of work through the transfer of working fluid; the Stirling concepts transfer work through a physical link between segments, such as a piston or rod. *Pressure ratio* identifies the concept's maximum to minimum pressure levels with respect to the other concepts considered. *Specific capacity* reflects useful heating or cooling capacity for a given volume. *External drive* indicates whether additional components, such as electric motors, are required to move the displacers. *Mechanical energy storage* indicates whether the concept requires some form of storage during its operating cycle. If storage is required, external drive is not required. For the concepts that require storage, it is achieved through the inertia of moving elements.

References

1. Pitcher, G. K., and F. K. du Pré, "Miniature Vuilleumier-Cycle Refrigerator," *Proceedings of the 1969 Cryogenic Engineering Conference*, pp. 447–451 (1970).
2. Finkelstein, T., "Balanced Compounding of Stirling Machines," paper 789194 presented at the 13th Intersociety Energy Conversion Engineering Conference, San Diego, August 1978.
3. Finkelstein, T., "Analysis of a Heat-Activated Stirling Heat Pump," paper No. 809363 presented at the 15th Intersociety Energy Conversion Engineering Conference, Seattle, Aug. 18–22, 1980.
4. Benson, G. M., "Free-Piston Heat Pumps," paper presented at the Institute of Gas Technology Stirling-Cycle Prime Movers Seminar, Rosemont, Ill., June 14–15, 1978. Published in *Seminar Proceedings*, pp. 67–99 (October 1979).

Chapter 6

Analytical Methodology

This chapter builds upon the basics discussed in Chaps. 3 and 4, and explains how the analysis of ideal cycles is brought closer to a realistic description of integrated heat pump performance. It reviews the analysis methods developed by others and describes, in general terms, the analysis methods used to calculate the performance comparisons of Chap. 7. The assumptions and typical operating conditions used in the analysis are discussed.

Stirling-Cycle Analysis

Most developed Stirling-cycle machines have followed the basic configuration of Robert Stirling, who conceived and pursued the development of an engine with two variable-volume working spaces operating on a regenerative thermodynamic cycle with two temperature levels. The ideal representation of these elements is shown in Fig. 6.1. This configuration is often used in describing the ideal Stirling engine and refrigerator cycles. Because of its simplicity, most of the analytical techniques that have been developed and corresponding published

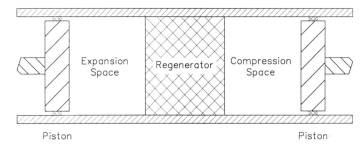

Figure 6.1 Basic Stirling-cycle configuration.

analyses are based on this configuration (with a few notable exceptions, such as *Mathematical Analysis of a Vuilleumier Refrigerator*[1]).

The analysis techniques are grouped into four main orders of complexity: zero, first, second, and third. Improvements in the basic understanding of the cycle and its thermodynamics and in the analytical and computational tools available have led to the more complex types of analysis.

Zero-order analysis assumes that the Stirling cycle is ideal and consists of four steps: isothermal expansion, constant-volume expansion with heat transfer, isothermal compression, and constant-volume expansion with heat transfer back to the initial condition. This analysis is limited to idealized investigations (see Chaps. 2 and 3) because of its assumptions:

- All the working fluid in a compression or expansion space is in the same thermodynamic state at any instant of time.
- The pressure is uniform throughout the working space at any instant.
- All other spaces in the machine, including the heat exchangers and regenerators, have zero gas volume, i.e., there is zero dead space.
- All the working fluid completely moves between the expansion and compression spaces (consistent with the zero-dead-space assumption).
- All moving parts move discontinuously from state to state.
- The compression and expansion steps take place under isothermal or isochoric conditions.
- The regenerator and heat exchangers are perfectly efficient. Temperature differences are not required to produce heat transfer, and the regenerator has no heat conduction throughout its mass.

Zero-order analysis treats the machine as an ideal cycle and provides an efficiency or coefficient of performance determination equivalent to that of a Carnot cycle. Thus, this ideal analysis overpredicts the capacity and efficiency of realistic Stirling engines and refrigerators. The cycle analysis can be described with closed-form analytical equations. This ideal analysis is useful only for comparative elementary studies and for establishing standard reference values for performance evaluation.

First-order analysis techniques are based on a classical analysis of Stirling engines that was introduced by G. Schmidt[2] in 1871. This analysis, based on what has become known as the Schmidt cycle, introduced harmonic motion of the pistons and displacers as a replacement for discontinuous motion. It takes into account the possibility of working fluid being in both the expansion and compression spaces during operation, thus allowing for part of the working fluid to be at each of two temperatures. It retains the other assumptions of the ideal analysis: zero dead-space volume, uniform pressure, perfect heat regeneration at constant volume, and isothermal expansion and compression. The Schmidt-cycle method predicts capacities of about 60 percent of that predicted by the ideal zero-order Stirling-cycle analysis, or less. The predicted thermal efficiency of Stirling engines and the coefficient of performance of Stirling refrigerators are again equivalent to those of a Carnot cycle and are not realistic.[3] As with the ideal analysis, a desirable feature of the Schmidt-cycle analysis is that the equations for work and heat transfer can be written in closed form and solved analytically.

The next advance in the analysis of Stirling machines was introduced by Dr. T. Finkelstein[4] in 1960. It is still classified as a first-order method because it makes the same assumptions of perfect regeneration and isothermal conditions in heat exchangers as the Schmidt-cycle analysis. However, Finkelstein's approach allows the expansion and compression in the working chambers to be described as anywhere from the one extreme of adiabatic processes to the other extreme of isothermal processes. In the limiting case of isothermal expansion and compression, the analysis corresponds to the Schmidt cycle.

Many people have used Finkelstein's adiabatic-cycle analysis to investigate the effect of varying the primary design parameters. In 1962 M. Khan[5] detailed the numerical results of his analysis, and Walker and Khan[6] summarized some of these results in 1965.

In 1968 E. B. Qvale and J. L. Smith[7] used an adiabatic cylinder analysis of the type that Finkelstein developed. They went further and developed a general set of equations for the basic analysis of what they termed a common generalized engine. Their analysis accounted for losses due to flow friction of the working fluid and blow-by past the pistons. It also allowed for finite heat-transfer rates by applying correction factors to the basic performance of the engine. Their performance predictions agreed well with tests of an engine developed for a space-power program by the Allison Division of General Motors Corporation. Their examples showed an agreement within about 7 percent of the heat input and work output and about 10 percent of the "internal efficiency," an efficiency Qvale and Smith calculated by correcting the power output to eliminate mechanical effects such as friction of the drive.

In addition to providing a reasonably accurate model, Qvale and Smith observed that the basic components can be arranged in different ways, but can be reduced to a single generalized engine. The engine geometry they referred to is the two-piston configuration shown in Fig. 6.1.

In summary, the above treatments are generally classified as first-order analysis techniques because they are based on the foundation that Schmidt provided.

Second-order analysis is represented by a number of different approaches. They generally build on the Schmidt analysis (or an equivalent approach), extrinsically adding additional heat flows to the basic results and adjusting the operating temperatures to account for less-than-ideal heat transfer.[8] Stirling experts differ on exactly how to classify these techniques, but they usually consider them somewhere between first- and third-order.

Second-order techniques usually use sets of equations that describe different aspects of the Stirling machine's operation, with no attempt made to develop closed-form analytical expressions. Second-order analysis can be reduced to the equivalent of first-order analysis by introducing simplifying assumptions that effectively remove the more detailed calculation of losses.

Nodal analysis and related techniques are usually classed as *third-order analyses*. Nodal analysis is similar in concept and solution to finite-element analysis and is based on analysis of nodes, which are small volumes of the working fluid or elemental slices of the operating space. The cycle is divided into a number of steps. Heat and mass transferred between working-fluid nodes and between nodes and the outside world are calculated for each node of the cycle. Nodal analysis is based on the size of each cycle step being small enough that the fluid properties are effectively uniform throughout the node. The step must also be small enough to eliminate the possibility of an element of working fluid both entering and leaving a node during the same step. Typically the fluid and material properties and heat and mass transfer exchanges are iteratively solved for until each time step produces a stable, internally consistent solution set. The calculations are repeated through the time steps to obtain a time-history of the cycle's performance.

Nodal analysis became feasible only when computer technology had developed to perform the thousands of repetitive calculations required. This is in sharp contrast to prior work because the calculation of performance no longer depended on closed-form analytical solutions involving difficult integrals.

The advantage of nodal analysis is its potential high accuracy. The time steps and number of nodes can be tailored to the accuracy needed. If the nodes and time steps are small enough and the heat and mass transfer functions are accurate, the resulting simulation can be very precise. The potential disadvantages are the relatively long calculation time (the set of calculations must be done for each time step and node) and the detailed understanding of the design required to define the calculations.

The most complete nodal analyses require a full specification of the design of the engine or refrigerator, including mechanical arrangement, materials, physical dimensions, fluid properties, and operating conditions. Property equations for the working fluid and equations for the conservation of mass, energy, and momentum are written for each node. Equations describing heat conduction along walls and through elements (like the pistons, displacers, and connecting rods) are also included.

The first nodal analysis program applied to Stirling engines was an adaptation of the National Aeronautics and Space Administration Thermal Analysis Program (TAP), by Dr. Finkelstein.[9] Further refinement and development of the program followed. In 1975 Finkelstein[10] presented the theoretical basis for analysis and described the technique used to solve the differential equations describing an engine. A similar program was made accessible through the Cybernet computer network.[11]

Urieli, Rallis, and Berchowitz[12] developed a Stirling-engine nodal analysis program that incorporated the same techniques as Finkelstein's. Additionally, they attempted to include real effects that had been omitted. Specifically, they noted that the momentum equations were for the steady-state case only, that kinetic-energy terms had not been included in the energy equations, and that the minimum number of nodes for the accurate simulation of a machine had not been specified. Their computer model corrected for the simplifications and resulting inaccuracies of earlier models. Their model predictions compared favorably with the results from a Stirling test engine that they had built. The computer program upon which this was based is more fully detailed in Urieli's Ph.D. thesis.[13]

Gedeon, then of Sunpower, Inc., in Athens, Ohio, outlined the techniques Sunpower used to simulate free-piston Stirling engines.[14] Sunpower has developed many tools for analyzing Stirling machines, especially free-piston units. These tools include a numerical simulation computer program for basic analysis and optimization of these engines. This simulation program allows the user to select the desired analysis complexity, from isothermal heat exchange or Schmidt cycles to complete nodal analysis, by specifying the appropriate input parameters. The program[15] is often used by Sunpower to aid in understanding and optimizing the Stirling engines under development, because of its ability to vary the depth of the analysis.

Schock[16] described the Stirling Nodal Analysis Program (SNAP) developed by Fairchild Industries under contract to the U.S. Department of Energy. The program was designed to rigorously analyze Stirling-cycle devices in conjunction with a parallel contract with Mechanical Technology, Inc., to develop a free-piston Stirling engine. Like the other nodal analysis programs described above, SNAP requires a division of the embodiment into individual nodes, which are numerically analyzed to determine the steady-state performance of the Stirling device. It is generally applicable to Stirling cycles, including free-piston configurations. It evaluates the motion of the piston due to gas and inertial forces and includes the effects of bounce spaces (gas springs), such as those found in Sunpower's free-piston Stirlings.

Recent Analytical Developments

The techniques described above represent the development of Stirling-machine analysis. The most recent work in the field has increased their accuracy by making fewer assumptions and developing numerical methods that are easier to use and more efficient.

Rauch[17] has described the "Harmonic Stirling Cycle Analysis Code" developed by Mechanical Technology, Inc. (MTI). MTI's approach to analyzing Stirling-cycle machines sought to reduce the long computation times required by the original, more detailed nodal analysis programs. It treats the thermodynamic space of the major classes of engine type—alpha (two pistons), beta (displacer and piston in same cylinder), and gamma (displacer and piston in separate cylinders)—by reducing them to an equivalent alpha configuration. This equivalence was also a conclusion reached by Qvale and Smith in 1968. Rauch assumed that the state variables can be adequately described by harmonic functions of time requiring only a few terms in the equation. With this assumption, the equations describing continuity, momentum, and energy may be solved analytically instead of numerically. The model that was developed under this assumption has been compared with the United Stirling P-40-7 and the GPU-3 engines for validation. The predicted power agreed with the measured values within 20 percent and the efficiency within 10 percent, after adjustment of the test data to account for drive and combustor losses and use of a multiplier. This multiplier corrected for the use of steady-state friction factors for periodic flow, entrance and exit losses, and other effects.

Another example of the ongoing improvements is Gedeon's[14] work on the Sunpower model. This program can simulate a Stirling engine which has six control volumes: one expansion space, one heater space, two regenerator spaces, one cooler space, and one compression space. Gedeon presented a method for reducing the number of iterations required to arrive at a stable numerical solution. The results for one cycle for the engine are calculated, and then the average temperatures required to achieve equilibrium energy flows are determined. This technique can approach equilibrium conditions faster than when the starting temperatures taken for the next cycle are those at the end of the previous cycle.

Urieli[18] has been working on a model with goals similar to Gedeon's. The intended attributes of this model are general utility, portability, maintainability, ease of use, and readability. Urieli has been developing a series of subroutines that may be combined to simulate a variety of Stirling-engine concepts. Three configurations were included: rhombic drive, swash-plate drive, and gamma (displacer and piston in separate cylinders). For actual analysis, the various drives are reduced to a standard equivalent configuration. Urieli has presented, along with the basic concept of an overall usable model, a method of reducing the number of nodes required to simulate regenerators. By using a least-squares fit to describe the temperature distribution in the regenerator, he has been able to reduce the required number of nodes to two, a very important analytical improvement.

In another successful attempt at simplification, Gedeon[19] produced a different technique and associated software package for developing solutions for the variables in nodal analysis. Instead of iterating through a number of time steps and calculating the nodal terms sequentially, Gedeon used Newton's method for solving systems of simultaneous linear equations to solve the time- and space-varying functions of the gas dynamic equations which describe a Stirling cycle. By simultaneously solving for all points using standard numerical methods, the conditions at any point and at any time are reached much more quickly. His approach also has the advantage of not needing the large number of time steps or nodes usually required for such analyses.

The analytical techniques described above apply to most single-cycle embodiments. As the concepts that combine engine and refrigerator cycles have attracted interest, the basic analytical techniques have been applied to them. Recent activity is shown in the works by Kinast,[20] Kuehl,[21] and Eder,[22] all of which applies to thermal-compression Vuilleumier concepts.

Researchers are continuing to refine their techniques for modeling Stirling cycles and related equipment. Additional effort is being devoted to testing and modeling individual components. These improvements can then be incorporated into overall models for detailed study, or their results can be used to refine the general calculations used in the more comprehensive models. Development of a computational tool that can determine the detailed performance of a design without resorting to repeated construction, testing, and refinement of prototype equipment would be a significant contribution.

Techniques Used in This Analysis

The goal of this book is the comparative analysis of Stirling, Vuilleumier, and Ericsson heat pump concepts, not the detailed design of any one embodiment. We suggest an analysis that includes enough details to identify important performance differences, yet does not require a detailed equipment design. An analysis that specifically treats one concept, and does not cover other concepts equally well, will give the same degree of uneven comparison that a detailed design approach would produce.

The technique selected for this analysis is categorized as second-order. Zero-order ideal analysis would result in the same performance predictions for each concept. It therefore has no value for comparative analysis. Analyzing the various concepts by first-order techniques would require the development and solution of complex equations describing the volume and pressure changes. However, implementing a numerical model to describe volume changes and calculate pressure changes and the resulting mass and heat transfer avoids (1) the possibility of incorrectly approximating the overall process and (2) the possibility of attempting to evaluate complex integrals which may have no solution or take an inordinate amount of time to solve. To produce equivalent simulation results, the techniques of nodal analysis were combined with first-order assumptions. To include the behavior of real equipment, the major losses that occur in such equipment were considered one at a time, so that their effect could be independently determined.

The basic model consists of the equations for two or more volumes, describing their variation as a function of a cycle step. To evaluate the concept's performance, the equations were stepped through one cycle to calculate a mass distribution which establishes a uniform pressure throughout. If the pressure and volume changes as a function of cycle position (or phase angle) are known, the heat and mass transfer and work produced or consumed can be calculated. The accuracy of this procedure has been confirmed by calculating the performance of an ideal piston-piston Stirling and comparing it with the results of a Schmidt analysis. (An analysis technique of a given order should, in general, be able to calculate the results of a lower order when the assumptions of the lower order are used.) Another confirmation of the general accuracy is that the calculated overall heat transferred at each of the temperature levels for the different concepts was the same.

Losses and nonideal effects are discussed in Chap. 7. The technique used to obtain the results presented in this book is general enough to be used equally for both the power-producing and power-consuming cycles.

Definition of Operating Conditions

To appropriately compare the various concepts under study, one of the first steps was to establish a common set of operating conditions that would be compatible with all the concepts. Various embodiments based on the Stirling cycle operate over a wide range of temperatures and temperature differences. Stirling engines have been designed to operate with hot-end working-fluid temperatures of over 750°C (1380°F), which corresponds to the use of high-temperature combustion products and other means of heating the working fluid. Engines have also been built that can operate with hot-end temperatures as low as about 50°C (120°F). The cold end of Stirling engines is typically water-cooled (as are automotive and stationary internal combustion engines) and usually ranges from just above 0°C (32°F) to 200°C (400°F). Temperatures even beyond these wide ranges have been suggested for non-heat pumping applications. The temperatures of the refrigerator segment of either a Vuilleumier or a Stirling-Stirling heat pump are dictated more by the outdoor and indoor air temperatures and are less under the control of a designer than the hot-end temperature of the engine segment.

To qualify our selection of operating conditions, we reviewed the conditions used in other Stirling and Vuilleumier systems. The highest heater head temperatures, around 750°C (1380°F), were usually associated with Stirling machines designed for high specific power or high engine efficiency. Increasing the hot-end temperature while maintaining the cold-end temperature increases the engine's temperature differential and its ideal efficiency. Lower hot-end temperatures allow more relaxed materials requirements. Such designs are more amenable to the use of traditional materials and construction techniques (such as steel instead of, for example, expensive high-temperature alloys or advanced ceramics).

Mean operating pressures also vary widely in these machines, from lows of 0.1 MPa (1 bar, 14.5 psia) to the maximum mean pressure that was observed in the literature, 15 MPa (2200 psia). It is theoretically possible to operate at even higher pressures, but the strength of the component materials is a limiting factor. These high pressure levels, like the high temperature levels, are associated with high-specific-power or high-efficiency machines. The operating limits selected for this analysis were not so extreme. Lower pressures would probably be used in comfort-heating applications because of cost and safety considerations. Machines not designed for high specific-power levels usually operate in the range of 3 to 6 MPa (30 to 60 bar, 435 psia to 870 psia). These lower pressures allow the use of simpler manufacturing techniques and more practical and affordable materials.

The operating temperatures and pressures are within the same range for Stirling-Stirling and Vuilleumier machines. The problems and solutions for high temperatures and pressures are the same for all cycles addressed in this book. However, the operating speeds vary more significantly because of the fundamental differences in how the cycles operate and practical limitations. These differences are discussed further below.

The speed ranges that are being used with current Stirling-technology machines are from 12.5 to 60 Hz (750 to 3600 rpm). Again, as with the higher temperatures and pressures, the higher speeds (30 to 60 Hz) are generally associated with high-specific-power engines. Increasing the speed of an engine is an easy method of increasing power output (as long as other limitations are not reached, such as the inability to transfer enough heat). If the work per cycle has been maximized, then increasing the number of cycles per minute will increase the power output. Free-piston machines also require higher speeds. Their operating speeds are determined by the natural frequency of the piston masses and gas or mechanical spring forces. Because of the size of the masses and leakage around the moving elements (which plays an increasing effect at lower speeds), free-piston engines and refrigerators usually operate in the 30- to 60-Hz range.

Low-speed operation also has disadvantages. Lubrication of moving parts can be more difficult. Heat-transfer coefficients are lower at lower working-fluid flow rates, and energy loss related to heat conduction between spaces becomes a more significant factor.

The speed range for thermal-compression machines is generally lower than that for mechanical-compression machines, dictated primarily by auxiliary power requirements. Thermal-compression machines require larger volumes of working fluid to achieve the same capacity per cycle. This leads to larger pressure drops and frictional losses when these working-fluid volumes flow through the heat exchangers. Thermal-compression machines also use external drives to move the displacers, with faster operation requiring higher drive power.

Based on these observations and considerations, we have chosen the following operating conditions for purposes of comparison: a mean operating pressure of 5 MPa (725 psia), an upper temperature of 538°C (1000°F), and a speed of 17 Hz (1000 rpm). The cold and warm temperatures are determined by the heat pump's environment. In the heating mode, this environment is defined by a desired indoor delivered-air temperature above 44°C (110°F), and an outdoor temperature predominantly between −18°C (0°F) and 10°C (50°F). For cooling, the operating temperatures are defined by the need to deliver air between 4°C (40°F) and 16°C (60°F), and reject heat to the ambient above 38°C (100°F). As a reasonable compromise between these sets of conditions, the heat input low temperature selected was 0°C (32°F). To ensure that there is sufficient heat rejection driving force in both heating and cooling modes, the heat rejection temperature selected was 66°C (150°F).

References

1. Sherman, A., *Mathematical Analysis of a Vuilleumier Refrigerator,* Report NASA-TM-X65534, X-763-71-125, Springfield, Va.: National Technical Information Center, 1971.
2. Schmidt, G., "Theorie der Lehmannscher Calorischen Maschine," *Z. Verb. dt. Ing., V15*(1) (1871).
3. Walker, G., *Stirling Engines,* London: Oxford University Press, 1980, pp. 50–51.
4. Finkelstein, T., "Generalized Thermodynamic Analysis of Stirling Engines," *SAE Paper 118B,* January 1960.
5. Khan, M. I., "The Application of Computer Techniques to the General Analysis of the Stirling Cycle," *Thesis,* Durham University, 1962.
6. Walker, G., and M. Khan, "The Theoretical Performance of Stirling Cycle Machines," *SAE Paper No. 949A,* Int. Automobile Engineering Congress, Detroit, 1965.
7. Qvale, E. B., and J. L. Smith, Jr., "A Mathematical Model for Steady Operation of Stirling-Type Engines," *Trans. ASME, J. Eng. Power,* 45–50 (January 1968).
8. Martini, W. R., and B. A. Ross, "An Isothermal Second-Order Stirling Engine Calculation Method," paper 799237 presented at the 14th Intersociety Energy Conversion Engineering Conference, Boston, Mass., 1979.
9. Finkelstein, T., G. Walker., and J. Joschi, "Design Optimization of Stirling-Cycle Cryogenic Cooling Engines, by Digital Simulation," paper K4, Cryogenic Engineering Conference, Boulder, Colo., June 1970.
10. Finkelstein, T., "Computer Analysis of Stirling Engines," paper 759140 presented at the 10th Intersociety Energy Conversion Engineering Conference, Newark, Del., 1975.
11. Walker, G., *Stirling Engines,* London: Oxford University Press, 1980, pp. 66–67.
12. Urieli, I., C. J. Rallis, and D. M. Berchowitz, "Computer Simulation of Stirling Cycle Machines," paper No. 779252 presented at the 12th Annual Intersociety Energy Conversion Engineering Conference, Washington, D.C., Aug. 28–Sept. 2, 1977, pp. 1512–1521.
13. Urieli, I., *A Computer Simulation of Stirling Cycle Machines,* Ph.D. Thesis, University of Witwatersrand, Johannesburg, South Africa, 1977.
14. Gedeon, D. R., "The Optimization of Stirling Cycle Machines," paper presented at the 13th Annual Intersociety Energy Conversion Engineering Conference, San Diego, August 1978.
15. Gedeon, D. R., lecture notes presented at A Stirling Engine Workshop by Ohio University and Sunpower, Inc., Athens, Ohio, Oct. 27–29, 1980.
16. Schock, A., "Nodal Analysis of Stirling Cycle Devices," paper presented at the 13th Annual Intersociety Energy Conversion Engineering Conference, San Diego, August 1978.
17. Rauch, J. S., "Harmonic Analysis of Stirling Engine Thermodynamics," paper

809335 presented at the 15th Annual Intersociety Energy Conversion Engineering Conference, Seattle, Aug. 18–22, 1980.
18. Urieli, I., "A General Purpose Program for Stirling Engine Simulation," paper 809336 presented at the 15th Annual Intersociety Energy Conversion Engineering Conference, Seattle, Aug. 18–22, 1980.
19. Gedeon, D., "A Globally-Implicit Stirling Cycle Simulation," paper 869121 presented at the 21st Intersociety Energy Conversion Engineering Conference, San Diego, Aug. 25–29, 1986.
20. Kinast, J. A., and J. Wurm, "Combined-Cycle Heat Pump Analysis Program," final report for Gas Research Institute under contract 5080-362-0355, December 1982.
21. Kuehl, H. D., N. Richter, and S. Schulz, "Computer Simulation of a Vuilleumier Cycle Heat Pump for Domestic Use," paper 869125 presented at the 21st Intersociety Energy Conversion Engineering Conference, San Diego, Aug. 25–29, 1986.
22. Eder, F. X., J. Blumenberg, W. Becker, M. Neubronner, A. Stübner, W. Messerschmidt, and A. Müller, "Der Vuilleumier-Prozess als Wärmepumpe und Kältemaschine; Analytische Behandlung und Messergebnisse (the Vuilleumier process as heat pump and cooling machine: analytical treatment and experimental results)," presented at the DKV 1989 Annual Meeting, Hannover, FRG, Nov. 22–24 (in German).

Chapter

7

Comparisons of Stirling-Stirling and Vuilleumier Heat Pumps

This chapter compares and assesses the operation and performance of Stirling-Stirling and Vuilleumier heat pumps. The six concept embodiments described in Chap. 5 are evaluated for nominal loads of 35-kW (10-ton) cooling capacity, and their relative merits are discussed.

The following discussion treats the concept embodiments in a different order from that in which they were presented in Chap. 5. As presented there, the order represented a rough measure of the concept embodiment's physical complexity, starting with the traditional Vuilleumier and moving toward multiple-cycle concepts. The order in this chapter corresponds to the complexity needed for analysis, moving from simpler analytical representations toward more complex ones. The thermal-compression Vuilleumier heat pumps are still first. The balanced-compounded Stirling machine follows next, preceding the duplex Stirling machine, because it can be analytically represented more simply. The balanced-compounded Vuilleumier and Ericsson-Ericsson heat pumps complete the analysis.

General Analysis Method

The concept embodiments are analyzed with second-order methods, as described in Chap. 6. Each concept embodiment is evaluated as an ideal cycle with an ideal gas working fluid. This means that the COPs and heating and cooling capacities are equal, but the size and capacity parameters differ. The general analysis calculates these differences.

Nonideal effects are considered and described later. This approach gives a better understanding of the processes involved than a first-order analysis, and yet is simpler and less tied to specific equipment design details than a third-order, or nodal, analysis.

Assumptions

Four basic assumptions underlie the analysis of the six concept embodiments presented in Chap. 5.

1. During the cycle, all working fluid in a space is at the space's temperature; that is, each space is isothermal. This isothermal assumption is commonly used in analyzing Stirling concepts. It requires the working fluid to exchange heat with the environment while in the space, not just in the heat exchangers. This does not describe the actual situation, in which heat is exchanged with the environment at the heat exchangers, while the working fluid in the spaces undergoes processes that are closer to adiabatic than to isothermal. However, for a comparative analysis, the error introduced by this assumption is small.

2. The heat exchangers and regenerators have zero volume, pistons and displacers traverse their entire cylinder space, and the fluid-flow passages connecting the engine and refrigerator spaces have negligible volume. This assumption of no dead space means that the working volume is totally devoted to the volume changes of the cycle. (For computational reasons, the simulation used a dead-space volume fraction of 0.001 to represent the no-dead-space condition.)

3. Pistons and displacers move in a sinusoidal pattern. Although free-piston configurations vary somewhat from this pattern, developers of these machines generally accept sinusoidal motion as a good approximation (as noted in Chap. 6).

4. The pressure is uniform at any instant throughout the working fluid in connected spaces. Working fluid separated by moving or fixed elements can have different pressures. In a real embodiment, the flow restrictions of connecting passages, heat exchangers, and regenerators and entrance and exit effects produce pressure differences between connected spaces. However, the pressure variation effects are usually not included until the later stages of equipment design.

Analytical simulation

We wrote a numerical model for each concept embodiment to describe the volume changes as a function of the cycle phase angle. The codes for these models are contained in Appendices A to E. In a second-order analysis, the traditional Vuilleumier and the Vuilleumier with internal heat exchangers are equivalent. The models computed the working-fluid distribution at any instant, for isothermal conditions in a space and uniform pressure throughout connected spaces. Perfect regenerator operation was simulated by having the working fluid leave and enter the spaces at the temperature corresponding to the space. The models determine the heat required to maintain reversible isothermal conditions in each of the chambers while pistons and/or displacers move. The models calculate the amount of heat absorbed and rejected by the working fluid for one complete cycle by evaluating multiple states along the cycle. The capacity per cycle times the operating speed yields the capacity per unit time. Once the heat requirement and the operating speed are known, the working volume required to meet the design capacity is calculated. The models also report the maximum and minimum volumes and working-fluid amounts residing in each space. These are used to evaluate some nonideal effects.

The values for volume and heat transfer from the analysis are reported to an artificially high precision to maintain the visibility of small differences between large numbers. It is understood that the results of testing a real machine would not approach this precision. However, this precision is used in the analysis to properly determine the net heat flows. In real equipment, only net heat flows would be measured experimentally, and the intermediate values shown for our analysis would not be seen.

Common operating conditions

As discussed in Chap. 6, selection of pressure, temperature, and speed parameters that apply to all the concept embodiments is important to enable a direct comparison. We selected a mean operating pressure of 5 MPa (750 psia); high, intermediate, and low working-fluid temperatures of 538°C (1000°F), 66°C (150°F), and 0°C (32°F), respectively; and an operating speed of 1000 rpm (16.7 Hz). The selected working fluid is helium.

The cooling capacity selected, 35 kW (10 tons; 120,000 Btu/h), represents a common commercial space conditioning equipment size. For capacities larger than 35 kW, multiple-cycle embodiments would be used. For the ideal analysis, this size results in a heating capacity of 49.5 kW (170,000 Btu/h). In real equipment, the ratio of heating capacity to cooling capacity would be greater because of regenerator inefficiencies and dead space. This is discussed later in this chapter when real equipment effects are considered.

After the ideal analysis, we parametrically evaluated nonideal effects to compare results among embodiments. For all embodiments, the heat exchangers operate similarly. The heat-exchanger volume may be approximately the same for each embodiment, but its proportion of the total working volume will differ. This chapter does not attempt to accurately quantify these effects as needed for detailed equipment design.

Ideal-Cycle Analysis

In our comparative analysis we identify the ideal (isentropic) operating characteristics of the various heat pump embodiments. The six embodiments described in Chap. 5—traditional Vuilleumier, Vuilleumier with internal heat exchangers, balanced-compounded Stirling, duplex Stirling, balanced-compounded Vuilleumier, and Ericsson-Ericsson—correspond to the six concepts: thermal-compression Vuilleumier, Vuilleumier with internal heat exchangers, piston-piston Stirling-Stirling, displacer-piston Stirling-Stirling, mechanical-compression Vuilleumier, and Ericsson-Ericsson, respectively. The first two embodiments are both thermal-compression concepts and are analytically equivalent with second-order techniques.

The four-cycle balanced-compounded heat pumps described in Chap. 5 have been modeled as a single cycle. They can be fully represented by analyzing one of the four identical cycles. Each cycle of a four-cycle embodiment will have one-fourth the working volume, capacity, and heat requirements given by the single-cycle analysis.

Thermal-compression Vuilleumier concepts (traditional Vuilleumier heat pump and Vuilleumier heat pump with internal heat exchangers embodiments)

The traditional Vuilleumier and Vuilleumier heat pump with internal heat exchangers embodiments are based on thermal-compression cycles. The pressure increase (compression) is achieved only by raising the average temperature of the working fluid. During operation, the total volume devoted to the working fluid does not change. These Vuilleumier heat pumps have traditionally been single-cycle machines that use displacers to move the working fluid between three different temperature regions. Our analysis did not differentiate heat-exchanger location and did not account for the differences in heat-exchange surface. Thus, the following discussion applies to both of these thermal-compression Vuilleumier concepts and the embodiments presented in Chap. 5.

When the engine and refrigerator are separate, there is only one engine-to-refrigerator volume ratio that will allow the engine to produce exactly the amount of work required to drive the refrigerator. This is because the input and output temperatures of both the engine and refrigerator are fixed. The work is transferred between the engine and refrigerator across some moving element like a piston. However, a thermal-compression Vuilleumier machine (with working fluid at any of three temperature levels) can operate with a range of engine-to-refrigerator volume ratios because the work transfer is done by the working fluid distributing itself between the hot, warm, and cold spaces. If the ratio is biased toward one space, the working fluid is distributed differently and the pressure ratio changes. As a result, the engine segment does not deliver one set amount of work; it delivers a range of work based on its pressure ratio. Similarly, the refrigerator can absorb a range of work amounts per cycle based on its pressure ratio. If the engine segment is very large with respect to the refrigerator segment, a large pressure ratio occurs. However, since there is little working fluid in the refrigerator, the engine can deliver very little work to drive the refrigerator, and therefore a large overall volume is needed to achieve a given design capacity. If the engine segment is very small with respect to the refrigerator segment, the low pressure ratio and small amount of working fluid in the engine segment also means that a large overall volume is required to achieve the design capacity. The minimum working volume occurs between these two extremes. The automatic balancing of engine and refrigerator capacities is accomplished by the working fluid movement. If the mean operating pressure in the engine segment is higher than that in the refrigerator segment, gas moves into the refrigerator from the engine. This increases the refrigerator segment's operating pressure and pressure ratio, and correspondingly reduces the engine's pressure, until the pressures are equal.

As these pressure ratios change, the capacity varies. To obtain the desired capacity, the overall size must be adjusted. To identify an optimum, we adjusted the engine-to-refrigerator volume ratio, for a design cooling capacity of 35 kW (10 tons), until the minimum total volume was found. The analysis results for that volume are shown in Table 7.1.

In general, the heat pump capacity depends on both the pressure ratio and the volume. For our second-order analysis, a concept embodiment's capacity is directly proportional to the refrigerator volume and is a monotonically increasing function of pressure ratio. In real machines this is not exactly true, because the effectiveness of the heat exchangers and regenerators also has an effect. However, for a fairly wide range of conditions, the relationship of pressure ratio and refrigerator volume with capacity is approximately as stated above; increasing either parameter increases the capacity. Our analysis shows that the thermal-compression Vuilleumier embodiments need a minimum

TABLE 7.1 Ideal Thermal-Compression Vuilleumier Heat Pump Analysis Results

Minimum Working Volume (Traditional Vuilleumier and Vuilleumier with Internal Heat Exchangers Embodiments)

Cooling capacity @ 1000 rpm	35 kW
Heating capacity	49.5 kW
Total working volume	3299 cm^3
Dead-space volume fraction	0.001
Minimum pressure	4.11 MPa
Maximum pressure	6.09 MPa
Mean pressure	5.00 MPa
Pressure ratio	1.48
Engine-segment volume	1865 cm^3
Refrigerator-segment volume	1434 cm^3

	Heat Transferred per Cycle		
Space	In	Out	Net
High temperature	9680 J	8815 J	865 J In
Intermediate (w/high)	8815 J	9680 J	865 J Out
Intermediate (w/low)	6139 J	8239 J	2100 J Out
Low temperature	8239 J	6139 J	2100 J In

1 Joule = 0.000948 Btu; 1 kW = 3413 Btu/h; 1 cm^3 = 0.061 in^3; 1 MPa = 145 psia.

total working volume of 3299 cm^3 (201 in^3) to achieve a 35-kW (10-ton) cooling capacity.

Thermal-compression Vuilleumier machines inherently have low pressure ratios. Table 7.1 shows this, with the ideal thermal-compression Vuilleumier having a pressure ratio of 1.48 when it operates between the above-given temperatures.

As noted above, these Vuilleumier heat pumps can be designed with different engine-to-refrigerator volume ratios. The effect of a different volume ratio can be seen by comparing the values shown in Tables 7.1 and 7.2. The pressure ratio is larger for the 2:1 volume ratio case (Table 7.2), because more fluid than for the minimum volume ratio case is going through temperature swings between 538°C (1000°F) and 66°C (150°F). The ability of the Vuilleumier heat pump to operate with different ratios allows it to adjust to varying temperatures without significantly deviating in its operating characteristics. Thus, design of the thermal-compression Vuilleumier does not require an exact match between engine and refrigerator volumes. For the higher pressure ratio case with the same heat pump capacity, the size of the refrigerator is smaller. Correspondingly, the lower refrigerator volume requires a larger engine volume. The larger engine volume is needed to move a larger volume of fluid and create a larger pressure ratio.

Stirling-Stirling concepts

Stirling engines and refrigerators are mechanical-compression machines. They differ from thermal-compression Vuilleumier machines

TABLE 7.2 Ideal Thermal-Compression Vuilleumier Heat Pump Analysis Results

Engine/Refrigerator Volume Ratio 2:1 (Traditional Vuilleumier and Vuilleumier with Internal Heat Exchanger Embodiments)

Cooling capacity @ 1000 rpm	35 kW
Heating capacity	49.5 kW
Total working volume	3455 cm^3
Dead-space volume fraction	0.001
Minimum pressure	3.94 MPa
Maximum pressure	6.35 MPa
Mean pressure	5.00 MPa
Pressure ratio	1.61
Engine-segment volume	2303 cm^3
Refrigerator-segment volume	1152 cm^3

	Heat Transferred per Cycle		
Space	In	Out	Net
High temperature	11,838 J	10,973 J	865 J In
Intermediate (w/high)	10,973 J	11,838 J	865 J Out
Intermediate (w/low)	4,752 J	6,852 J	2,100 J Out
Low temperature	6,852 J	4,752 J	2,100 J In

1 Joule = 0.000948 Btu;, 1 kW = 3413 Btu/h; 1 cm^3 = 0.061 in^3; 1 MPa = 145 psia.

in that their pistons mechanically alter the working-fluid volume. This is why in this book Stirling concepts have been described as mechanical-compression machines.

Piston-piston Stirling-Stirling (balanced-compounded Stirling embodiment)
Table 7.3 shows the operating performance of the balanced-compounded Stirling embodiment of a piston-piston Stirling-Stirling heat pump. This embodiment consists of four identical cycles. As stated previously, our models simulate single-cycle embodiments. Thus, the calculated capacities and volumes for each cycle of the four-cycle balanced-compounded embodiment are one-fourth the values in Table 7.3. This analysis was similar to that for the thermal-compression Vuilleumier, but with two main differences. First, Stirling-Stirling concepts do not have fluid passages between the warm spaces of the engine and refrigerator segments. Therefore, the pressures in the Stirling-Stirling engine and refrigerator segments are not necessarily equal. The second difference is that the linked pistons transfer work to one another. Because this work is transferred between the engine and refrigerator segments during the cycle, the net work generated in the engine segment must equal the net work absorbed in the refrigerator segment. This was simulated by first calculating the work per cycle for each segment, then sizing the refrigerator segment based on the design capacity and cycling speed. Then the engine segment was sized based on the net work required by the refrigerator segment.

TABLE 7.3 Ideal Piston-Piston Stirling-Stirling Heat Pump Analysis Results
(Balanced-Compounded Stirling Embodiment)

Cooling capacity @ 1000 rpm		35 kW
Heating capacity		49.5 kW
Total working volume		1281 cm^3
Dead-space volume fraction		0.001

	Engine	Refrigerator
Minimum pressure	1.83 MPa	2.06 MPa
Maximum pressure	13.66 MPa	12.16 MPa
Mean pressure	5.00 MPa	5.00 MPa
Pressure ratio	7.46	5.90
Volume	258 cm^3	1023 cm^3

Heat Transferred per Cycle			
Space	In	Out	Net
High temperature—engine	1140 J	275 J	865 J In
Intermediate—engine	398 J	759 J	361 J Out
Intermediate—refrigerator	1323 J	3927 J	2604 J Out
Low temperature—refrigerator	3551 J	1451 J	2100 J In

1 Joule = 0.000948 Btu; 1 kW = 3413 Btu/h; 1 cm^3 = 0.061 in^3; 1 MPa = 145 psia.

Comparing the results in Tables 7.1 and 7.2 with those in Table 7.3 shows the differences between the lower-pressure-ratio thermal-compression Vuilleumier and the higher-pressure-ratio Stirling-Stirling machine. The piston-piston Stirling engine and Stirling refrigerator operate with pressure ratios of 7.45 and 5.90, respectively. The thermal-compression Vuilleumier with the minimum working volume has a pressure ratio of only 1.48.

This difference in pressure ratios results in a significant difference in working volumes needed to produce the same overall capacity. The combined volume of the piston-piston Stirling engine and refrigerator is only 1275 cm^3 (77.8 in^3), about 39 percent of the volume of the smallest Vuilleumier with the same capacity.

Displacer-piston Stirling-Stirling (duplex Stirling embodiment) The duplex Stirling heat pump discussed and illustrated in Chap. 5 is an embodiment of a displacer-piston Stirling-Stirling concept. The engine segment and the refrigerator segment each have one displacer, and share a common piston. (An alternative arrangement could use two pistons connected so that they operate as the two halves of the shared piston.) The engine displacer separates the hot and warm spaces. The refrigerator displacer separates the cold and warm spaces. The engine segment produces power and transfers it to the refrigerator segment through the shared piston. The piston is on the warm side of both the engine and the refrigerator so that the sealing components are not exposed to temperature extremes during operation.

The duplex Stirling does not achieve pressure ratios as high as those achieved by the balanced-compounded Stirling. This can be explained using two configurations called gamma and beta. Figure 7.1 depicts the gamma configuration, in which a displacer and piston are placed in two separate cylinders with a flow connection between them. In this configuration, the expansion space (hot space for an engine) is formed between the displacer and one end of the cylinder. The compression space is the sum of the space between the displacer and the other end of the cylinder and the space between the piston and its cylinder's end. The compression ratio that can be achieved is

$$\frac{\text{Displacer cylinder volume} + \text{piston cylinder maximum volume}}{\text{Displacer cylinder volume} + \text{piston cylinder minimum volume}}$$

The gamma configuration adds a space, which is not reached by the piston, to the operating space because the displacer cylinder volume is constant. The pressure ratio is reduced as a result of this additional space.

Figure 7.2 illustrates the beta configuration, in which the displacer and piston are in the same cylinder. (The duplex Stirling configuration is an example of a beta configuration.) The expansion space is the same as for the gamma configuration, but the compression space is formed between the displacer and the piston, without the cylinder

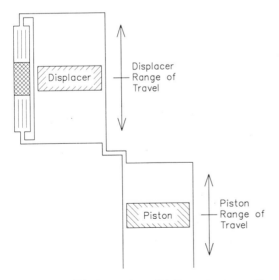

Figure 7.1 Displacer-piston Stirling using two cylinders (gamma configuration).

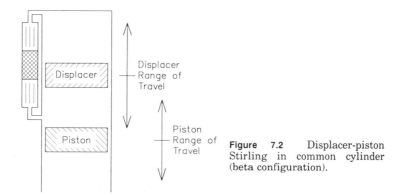

Figure 7.2 Displacer-piston Stirling in common cylinder (beta configuration).

ends and connections. If the ranges of travel for the displacer and the piston do not overlap, the compression ratio is the same as above. If the piston is brought closer to the displacer, so that their paths overlap, the amount of additional space is reduced and the compression ratio increases. However, the additional space cannot be reduced to zero because the movements of the piston and the displacer are out of phase with each other. If the ranges of travel are made to overlap such that the additional space becomes zero, the piston and displacer come into contact, perturbing their motion and potentially disrupting the machine's operation.

To prevent contact between the piston and displacer in the analysis, a small offset volume is added. This offset is just large enough so that the compression volume (total volume minus the expansion volume) is always positive.

The different physical arrangement of the displacer-piston embodiment compared with the piston-piston embodiment described in Chap. 5 results in a number of differences in operation. The pistons in a piston-piston embodiment travel in their own range; that is, their respective paths of travel do not overlap. Also, the ratio of maximum volume to minimum volume for the working fluid, which affects the pressure ratio, is smaller than in the displacer-piston configuration.

Table 7.4 shows the results of the analysis for the duplex Stirling embodiment of the displacer-piston Stirling-Stirling heat pump. The displacer-piston Stirling-Stirling machine needs 14 percent more total volume than the piston-piston Stirling-Stirling machine with the same capacity. Correspondingly, the refrigerator pressure ratio for the displacer-piston embodiment is lower. Because of this pressure ratio drop from 5.90 to 2.27, a greater working volume is required to achieve the same capacity, 1167 cm^3 (71.2 in^3) compared with 1023 cm^3 (62.4 in^3). This larger volume means that more heat exchange is required to maintain isothermal operation; that is, either more heat

TABLE 7.4 Ideal Displacer-Piston Stirling-Stirling Heat Pump Analysis Results
(Duplex Stirling Embodiment)

Cooling capacity @ 1000 rpm		35 kW
Heating capacity		49.5 kW
Total working volume		1497 cm^3
Dead-space volume fraction		0.001

	Engine	Refrigerator
Minimum pressure	2.38 MPa	3.32 MPa
Maximum pressure	10.51 MPa	7.52 MPa
Mean pressure	5.00 MPa	5.00 MPa
Pressure ratio	4.42	2.27
Volume	307 cm^3	1167 cm^3

	Heat Transferred per Cycle		
Space	In	Out	Net
High temperature—engine	1368 J	503 J	865 J In
Intermediate—engine	994 J	1355 J	361 J Out
Intermediate—refrigerator	3588 J	6192 J	2604 J Out
Low temperature—refrigerator	4547 J	2447 J	2100 J In

1 Joule = 0.000948 Btu; 1 kW = 3413 Btu/h; 1 cm^3 = 0.061 in^3; 1 MPa = 145 psia.

transfer surface or a higher heat transfer rate per unit area is required. Similarly, for the engine segments, the pressure ratios are lower, and working volumes and heat transfer requirements are higher for the displacer-piston Stirling-Stirling machine than for the piston-piston Stirling-Stirling machine.

Mechanical-compression Vuilleumier concept (balanced-compounded Vuilleumier embodiment)

The mechanical-compression Vuilleumier augments the thermal compression of the traditional Vuilleumier with mechanical compression. A mechanical-compression classification is not completely accurate because a portion of the working-fluid pressure change is caused by the shift of working fluid from a warm space to a hot space (as in thermal compression). However, this embodiment would not operate as only a thermal-compression machine (without altering the volume). Only a cyclic pressure variation would be produced as the gas moved between the two temperature levels, and no net work would be produced or consumed.

The working fluid in the mechanical-compression Vuilleumier concept comes in contact with three different temperature levels, as in all Vuilleumier concepts. However, instead of using displacers to shuttle gas between the spaces, pistons are used to move the working fluid and to compress and expand the gas mechanically. This gives the mechanical-compression Vuilleumier machines the potential for higher pressure ratios than thermal-compression Vuilleumier machines.

Unlike the thermal-compression Vuilleumier concepts and like the Stirling-Stirling concepts, the mechanical-compression Vuilleumier can produce or absorb work because it uses mechanical compression. The hot space and part of the warm space are equivalent to the engine segment of a Stirling-Stirling. The remaining portion of the warm space and the cold space are equivalent to the refrigerator segment. If the refrigerator segment requires less power than the engine segment delivers, either the temperatures shift to reach a new equilibrium point (producing greater heat pump capacity) or excess power is available. If the refrigerator segment requires more power than the engine segment delivers, either the temperatures shift (reducing the heat pump capacity) or power must be added. Our mechanical-compression Vuilleumier embodiment analysis adjusted the size of the cold space with respect to the hot and warm spaces until the net heat absorbed by the hot and cold spaces equaled the net heat rejected by the warm space. When this is done, the mechanical-compression Vuilleumier embodiment can be sized and compared with the other embodiments. The size of the cold space was varied with respect to the hot and warm spaces to reflect the design of the balanced-compounded Vuilleumier. The hot and warm spaces share the same cylinders, so their volumes are equal, while the cold spaces have their own cylinder, which can have a different diameter.

The results for the balanced-compounded Vuilleumier embodiment are shown in Table 7.5. This embodiment consists of four identical cycles, but as stated previously, our analysis simulates single-cycle embodiments. Thus, the results for each cycle of the four-cycle balanced-compounded embodiment are one-fourth the values in Table 7.5. A comparison of the heat transferred per cycle for this mechanical-compression Vuilleumier embodiment with that transferred by the thermal-compression Vuilleumier and the Stirling-Stirling embodiments shows the intermediate values of the mechanical-compression Vuilleumier. Like all Vuilleumier concepts, the three characteristic operating temperatures correspond to one set of maximum, minimum, and average pressures. However, the operational values are closer to those of Stirling-Stirling machines because both operate using mechanical compression. For the minimum working volume, the thermal-compression Vuilleumier analysis yielded a pressure ratio of 1.48 and the Stirling-Stirling analysis yielded pressure ratios of 7.46 (engine) and 5.90 (refrigerator), while the mechanical-compression Vuilleumier analysis yielded a pressure ratio of 9.13. The total work-

TABLE 7.5   Ideal Mechanical-Compression Vuilleumier Heat Pump Analysis Results
(Balanced-Compounded Vuilleumier Embodiment)

Cooling capacity @ 1000 rpm	35 kW
Heating capacity	49.5 kW
Total working volume	1334 cm^3
Dead-space volume fraction	0.001
Minimum pressure	1.64 MPa
Maximum pressure	15.24 MPa
Mean pressure	5.00 MPa
Pressure ratio	9.29
Cold volume/total volume	0.7082

	Heat Transferred per Cycle		
Space	In	Out	Net
High temperature	2090 J	1225 J	865 J In
Intermediate temperature	733 J	3698 J	2965 J Out
Low temperature	5073 J	2973 J	2100 J In

1 Joule = 0.000948 Btu; 1 kW = 3413 Btu/h; 1 cm^3 = 0.061 in^3; 1 MPa = 145 psia.

ing volumes also reflect this difference in operation: the thermal-compression Vuilleumier needs 3299 cm^3 (201.3 in^3), while the piston-piston Stirling-Stirling and mechanical-compression Vuilleumier have total volumes of 1281 cm^3 (78.2 in^3) and 1334 cm^3 (81.4 in^3), respectively.

Ericsson-Ericsson concept

As discussed in Chap. 5, the Ericsson-Ericsson heat pump is similar in principle to the balanced-compounded, mechanical-compression Vuilleumier. Analysis of the Ericsson-Ericsson embodiment requires the determination of five separate volumes. The hot- and cold-space volumes vary approximately sinusoidally as the displacers move. The other three volumes are warm spaces, two of which are on opposite sides of the displacers and are equal in size to, but 180° out of phase with, the hot and cold spaces. They are connected by flow passages to each other and to the third warm space, which is between the flywheel pistons. The ratio of the hot-space volume to the cold-space volume, and the fraction of the total volume devoted to the warm space between the pistons characterize the machine. The ratio between the hot and cold spaces determines whether net work is generated or absorbed by the cycle.

The Ericsson-Ericsson heat pump generates work during part of the cycle and absorbs it during another part. Energy is stored in the flywheel pistons as they accelerate and are acted upon by the working fluid. This energy is released again as the flywheel pistons slow down while compressing the working fluid. If the engine segment generates more energy than the refrigerator segment requires, the pistons will change their stroke. If the energy difference is too large, the pistons will hit the ends of their cylinders. Alternatively, if more power is required than is generated, the system will slow down or stop. The analysis showed that the ratio of cold- to hot-space volumes is about 2.4:1 for the temperatures we selected.

Although the Ericsson-Ericsson is like the thermal-compression Vuilleumier in that the working fluid can be associated with any one of three temperatures, its displacers move in phase with each other rather than with a 90° phase difference. The pressure variations that drive the cycle are instead generated by the changing volume between the flywheel pistons. If this interpiston volume is small relative to the total volume, very small pressure swings occur. If the interpiston volume is too large, little working fluid is present in the other spaces to absorb and reject heat. The fraction of the total working volume needed in the space between the pistons appears to be independent of the operating temperatures. Iterations with the model showed that about 62 percent of the total volume should be allocated to the interpiston space to produce the most heat absorbed at the low temperature per cycle. This ratio was used throughout the analysis.

Table 7.6 shows the results for the Ericsson-Ericsson embodiment. Comparing these results with those for the balanced-compounded Vuilleumier, its closest thermodynamic counterpart, shows the differences between the Ericsson-Ericsson and this mechanical-compression

TABLE 7.6 Ideal Ericsson-Ericsson Heat Pump Analysis Results

(Ericsson-Ericsson Embodiment)

Cooling capacity @ 1000 rpm	35 kW
Heating capacity	49.5 kW
Total working volume	2098 cm^3
Dead-space volume fraction	0.001
Minimum pressure	3.08 MPa
Maximum pressure	8.11 MPa
Mean pressure	5.00 MPa
Pressure ratio	2.63
Cold volume/hot volume	2.4274
Interpiston volume/total volume	0.62

Heat Transferred per Cycle			
Space	In	Out	Net
High temperature	1637 J	225 J	865 J In
Intermediate temperature	6010 J	8975 J	2965 J Out
Low temperature	3975 J	1875 J	2100 J In

1 Joule = 0.000948 Btu; 1 kW = 3413 Btu/h; 1 cm^3 = 0.061 in^3; 1 MPa = 145 psia.

Vuilleumier embodiment. The Ericsson-Ericsson embodiment volume requires almost 60 percent more working volume than the balanced-compounded Vuilleumier to produce the same capacity. The pressure ratio, 2.63 for the Ericsson-Ericsson compared with 9.13 for the balanced-compounded Vuilleumier, also reflects this larger size. Only a portion of the warm space in the Ericsson-Ericsson embodiment is directly acted upon by the mechanical compression of the flywheel pistons, whereas all three temperature spaces are mechanically compressed in the balanced-compounded Vuilleumier. This has the effect of adding dead space to an Ericsson-Ericsson machine, and results in a drop in pressure ratio and an increase in required volume.

Comparing the heat transferred during the cycle shows that the hot and cold heat exchangers are less heavily loaded than those of the balanced-compounded Vuilleumier. However, the warm heat exchangers are loaded more heavily. This is a direct result of the Ericsson-Ericsson having most of the working fluid in the warm space and in an isothermal state. Because the Ericsson-Ericsson embodiment does not have heat exchangers for the space between the flywheel pistons, this space will deviate from the isothermal heat transfer assumed in the analysis.

Effect of More Realistic Design Factors

Dead space

Dead space is the space which is not part of any of the cyclically varying volumes which produce the thermodynamic cycle. These spaces include connecting passages, heat exchangers, regenerators, and piston and displacer clearance spaces. Although most of these spaces are needed to produce a practical machine, they are not included in the ideal machine. The ideal machine implementing an ideal cycle works on the assumption that the heat transfer required to maintain an isothermal working fluid occurs through the surfaces of the cylinders which contain the working fluid. In reality, these surfaces are not large enough to permit the necessary heat transfer, and separate heat-exchange surfaces must be introduced to approach isothermal heat transfer. These additional heat exchangers do not participate in the volume changes, but they do contribute to the total working volume.

The additional heat-exchange surfaces require a working volume increase larger than just the added heat-exchange volume for the same capacity (as explained below). To achieve the same mean operating pressure, more working fluid must be introduced. This additional fluid increases the amount of heat that must be transferred to approach isothermal heat exchange. The dead-space volume contains fluid that does not directly participate in the volume and temperature changes, but acts as a buffer space. The pressure swings that occur are reduced because of this buffer space. Because the pressure ratio is reduced, the volume must be increased further to maintain the design capacity. The pressure ratio approaches 1 as the dead-space volume is increased to the limit of occupying the entire working space. At this limit, no working fluid is directly participating in the temperature and volume changes. With no pressure changes, no useful heating, cooling, or work is produced. This is why it is important to minimize the dead space in real equipment.

As an example of the effect of dead space, consider the cylinder in Fig. 7.3a, which has a 10:1 volume ratio (that is, its maximum volume is 10 times its minimum volume). If, as shown in Fig. 7.3b, 20 percent dead space is added, the resulting maximum volume is 12 times the initial minimum volume, while the new minimum volume is 3 times the original minimum. This produces a 12:3, or 4:1, volume ratio. This dead space substantially decreases the pressure ratio, and even though the dead space results in a larger working volume, if only the dead space is added, the heat pump's capacity will decrease. To achieve the same capacity (without operating condition changes), the working volume must be increased.

To simulate the effect of dead space, the models used in the ideal cycle analysis were modified so that the piston and/or displacer travel

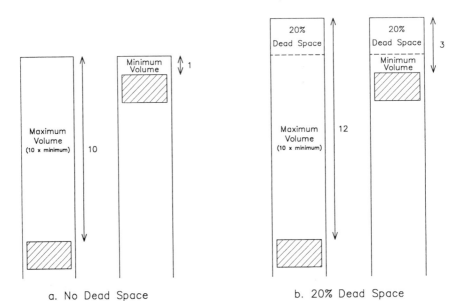

Figure 7.3 Effect of dead space on volume compression ratio.

always left a portion of the space unused. As a result, the dead space was distributed in the same ratio as the working-volume ratios, simulating a uniform distribution of dead space. In a real machine, the dead-space distribution would not be uniform, but would depend on the detailed design requirements. If additional heat-transfer surfaces are required to achieve the same heat transfer with smaller temperature differences, the dead-space volume will be increased where these surfaces are located. It is likely that dead space will be associated with the warm- and cold-space volumes because of the relatively small temperature differences present. Less of the dead space will be associated with the hot-space volume, because a reduced heat transfer area can be partially compensated for by increasing the firing temperature. However, for analytical comparison purposes, a uniform distribution of dead space is an adequate description. The working volume was sized to maintain the 35-kW (10-ton) design cooling capacity. The rest of the procedure was the same as for the ideal-cycle analysis.

Tables 7.7 and 7.8 show the working-volume requirements and pressure ratios for dead-space volume fractions of 0.001 (no dead space), 0.2, and 0.4, for the concept embodiments presented in this book. Table 7.7 shows that for all embodiments the working volume increases from 52 percent to 76 percent for a dead-space volume fraction of 0.2, and from 174 percent to 226 percent for a dead-space volume fraction of 0.4, as compared with the no-dead-space machine. Thus, the working-volume ratio for the embodiments is proportionally about the same and is approximately independent of the dead-space volume fraction. Table 7.8 shows that embodiments with higher no-dead-space pressure ratios have larger pressure-ratio reductions than embodiments with lower no-dead-space pressure ratios as the dead-space volume fraction is increased.

The performance of the thermal-compression Vuilleumier embodiments (traditional Vuilleumier and Vuilleumier heat pump with in-

TABLE 7.7 Working Volume for Dead-Space Volume Fractions of 0.001, 0.2, and 0.4

	Working volume, cm^3		
Dead-space volume fraction	0.001	0.2	0.4
Traditional Vuilleumier and Vuilleumier with internal heat exchangers	3299	5153	9185
Balanced-compounded Stirling	1281	2142	3987
Duplex Stirling	1474	2462	4597
Balanced-compounded Vuilleumier	1334	2034	4353
Ericsson-Ericsson	2098	3340	6029

1 cm^3 = 0.061 in^3.

TABLE 7.8 Pressure Ratios for Dead-Space Volume Fractions of 0.001, 0.2, and 0.4

Dead-space volume fraction	Pressure ratios		
	0.001	0.2	0.4
Traditional Vuilleumier and Vuilleumier with internal heat exchangers	1.48	1.37	1.26
Balanced-compounded Stirling—engine	7.46	4.16	2.69
—refrigerator	5.90	3.64	2.49
Duplex Stirling—engine	4.42	2.74	2.01
—refrigerator	2.27	1.86	1.65
Balanced-compounded Vuilleumier	9.29	4.64	2.87
Ericsson-Ericsson	2.63	2.12	1.74

ternal heat exchangers) is more tolerant of larger dead-space volume than that of the mechanical-compression embodiments. Although their working volume is higher, for the same heat pump capacity the thermal-compression Vuilleumier embodiments add proportionally less total volume to include the same volume of dead space. Also, because the dead-space volume fraction for a thermal-compression Vuilleumier translates to a larger absolute volume than for mechanical-compression concepts, the heat-exchanger and regenerator volumes (which constitute most of the dead space) can be made larger to improve their performance. These relatively large thermal-compression machines are not very sensitive to dead space. Embodiments of these concepts should either have a smaller dead-space volume fraction or improved performance as compared with the smaller mechanical-compression machines. Correspondingly, the smaller, higher-pressure-ratio mechanical-compression machines are more sensitive to an increase in dead space. These embodiments will need to be carefully designed to remain compact and to retain high performance.

Tables 7.9 to 7.18 show the analysis results for dead-space volume fractions of 0.2 and 0.4, for the thermal-compression Vuilleumier, piston-piston Stirling-Stirling, displacer-piston Stirling-Stirling, mechanical-compression Vuilleumier, and Ericsson-Ericsson embodiments, respectively.

Regenerator inefficiencies

One source of performance loss in integrated heat pumps is regenerator inefficiency. Regenerators are reservoirs in which heat is deposited and removed by the working fluid as it passes from a space at one temperature to a space at a different temperature. The regenerator allows the gas to enter the new space near the temperature of the gas already present in the space. If no regenerator were used, the working fluid

TABLE 7.9 Ideal Thermal-Compression Vuilleumier Heat Pump (with 20 percent Dead Space) Analysis Results

Minimum Working Volume Case (Traditional Vuilleumier and Vuilleumier with Internal Heat Exchangers Embodiments)

Cooling capacity @ 1000 rpm	35 kW
Heating capacity	49.5 kW
Total working volume	5153 cm^3
Dead-space volume fraction	0.2
Minimum pressure	4.27 MPa
Maximum pressure	5.85 MPa
Mean pressure	5.00 MPa
Pressure ratio	1.37
Engine-segment volume	2913 cm^3
Refrigerator-segment volume	2240 cm^3

	Heat Transferred per Cycle		
Space	In	Out	Net
High temperature	12,043 J	11,178 J	865 J In
Intermediate (w/high)	11,178 J	12,043 J	865 J Out
Intermediate (w/low)	7,941 J	10,041 J	2,100 J Out
Low temperature	10,041 J	7,941 J	2,100 J In

1 Joule = 0.000948 Btu; 1 kW = 3413 Btu/h; 1 cm^3 = 0.061 in^3; 1 MPa = 145 psia.

TABLE 7.10 Ideal Thermal-Compression Vuilleumier Heat Pump (with 40 percent Dead Space) Analysis Results

Minimum Working Volume Case (Traditional Vuilleumier and Vuilleumier with Internal Heat Exchangers Embodiments)

Cooling capacity @ 1000 rpm	35 kW
Heating capacity	49.5 kW
Total working volume	9185 cm^3
Dead-space volume fraction	0.4
Minimum pressure	4.45 MPa
Maximum pressure	5.62 MPa
Mean pressure	5.00 MPa
Pressure ratio	1.26
Engine-segment volume	5192 cm^3
Refrigerator-segment volume	3993 cm^3

	Heat Transferred per Cycle		
Space	In	Out	Net
High temperature	15,977 J	15,112 J	865 J In
Intermediate (w/high)	15,112 J	15,977 J	865 J Out
Intermediate (w/low)	10,952 J	13,052 J	2,100 J Out
Low temperature	13,052 J	10,952 J	2,100 J In

1 Joule = 0.000948 Btu; 1 kW = 3413 Btu/h; 1 cm^3 = 0.061 in^3; 1 MPa = 145 psia.

TABLE 7.11 Ideal Piston-Piston Stirling-Stirling Heat Pump (with 20 percent Dead Space) Analysis Results

(Balanced-Compounded Stirling Embodiment)

Cooling capacity @ 1000 rpm	35 kW
Heating capacity	49.5 kW
Total working volume	2142 cm^3
Dead-space volume fraction	0.2

	Engine	Refrigerator
Minimum pressure	2.45 MPa	2.62 MPa
Maximum pressure	10.18 MPa	9.54 MPa
Mean pressure	5.00 MPa	5.00 MPa
Pressure ratio	4.16	3.64
Volume	437 cm^3	1705 cm^3

	Heat Transferred per Cycle		
Space	In	Out	Net
High temperature—engine	1354 J	489 J	865 J In
Intermediate—engine	646 J	1007 J	361 J Out
Intermediate—refrigerator	2160 J	4763 J	2604 J Out
Low temperature—refrigerator	4418 J	2318 J	2100 J In

1 Joule = 0.000948 Btu; 1 kW = 3413 Btu/h; 1 cm^3 = 0.061 in^3; 1 MPa = 145 psia.

TABLE 7.12 Ideal Piston-Piston Stirling-Stirling Heat Pump (with 40 percent Dead Space) Analysis Results

(Balanced-Compounded Stirling Embodiment)

Cooling capacity @ 1000 rpm	35 kW
Heating capacity	49.5 kW
Total working volume	3987 cm^3
Dead-space volume fraction	0.4

	Engine	Refrigerator
Minimum pressure	3.05 MPa	3.17 MPa
Maximum pressure	8.21 MPa	7.89 MPa
Mean pressure	5.00 MPa	5.00 MPa
Pressure ratio	2.69	2.49
Volume	819 cm^3	3168 cm^3

	Heat Transferred per Cycle		
Space	In	Out	Net
High temperature—engine	1696 J	831 J	865 J In
Intermediate—engine	1015 J	1376 J	361 J Out
Intermediate—refrigerator	3486 J	6090 J	2604 J Out
Low temperature—refrigerator	5770 J	3670 J	2100 J In

1 Joule = 0.000948 Btu; 1 kW = 3413 Btu/h; 1 cm^3 = 0.061 in^3; 1 MPa = 145 psia.

TABLE 7.13 Ideal Displacer-Piston Stirling-Stirling Heat Pump (with 20 percent Dead Space) Analysis Results

(Duplex Stirling Embodiment)

Cooling capacity @ 1000 rpm		35 kW
Heating capacity		49.5 kW
Total working volume		2462 cm^3
Dead-space volume fraction		0.2

	Engine	Refrigerator
Minimum pressure	3.02 MPa	3.67 MPa
Maximum pressure	8.27 MPa	6.82 MPa
Mean pressure	5.00 MPa	5.00 MPa
Pressure ratio	2.74	1.86
Volume	633 cm^3	1829 cm^3

Heat Transferred per Cycle			
Space	In	Out	Net
High temperature—engine	1752 J	887 J	865 J In
Intermediate—engine	1584 J	1945 J	361 J Out
Intermediate—refrigerator	5058 J	7662 J	2604 J Out
Low temperature—refrigerator	5578 J	3478 J	2100 J In

1 Joule = 0.000948 Btu; 1 kW = 3413 Btu/h; 1 cm^3 = 0.061 in^3; 1 MPa = 145 psia.

TABLE 7.14 Ideal Displacer-Piston Stirling-Stirling Heat Pump (with 40 percent Dead Space) Analysis Results

(Duplex Stirling Embodiment)

Cooling capacity @ 1000 rpm		35 kW
Heating capacity		49.5 kW
Total working volume		4597 cm^3
Dead-space volume fraction		0.4

	Engine	Refrigerator
Minimum pressure	3.53 MPa	3.89 MPa
Maximum pressure	7.08 MPa	6.43 MPa
Mean pressure	5.00 MPa	5.00 MPa
Pressure ratio	2.01	1.65
Volume	1463 cm^3	3134 cm^3

Heat Transferred per Cycle			
Space	In	Out	Net
High temperature—engine	2,300 J	1,435 J	865 J In
Intermediate—engine	2,389 J	2,750 J	361 J Out
Intermediate—refrigerator	6,523 J	9,127 J	2,604 J Out
Low temperature—refrigerator	6,608 J	4,508 J	2,100 J In

1 Joule = 0.000948 Btu; 1 kW = 3413 Btu/h; 1 cm^3 = 0.061 in^3; 1 MPa = 145 psia.

TABLE 7.15 Ideal Mechanical-Compression Vuilleumier Heat Pump (with 20 percent Dead Space) Analysis Results

(Balanced-Compounded Vuilleumier Embodiment)

Cooling capacity @ 1000 rpm	35 kW
Heating capacity	49.5 kW
Total working volume	2034 cm^3
Dead-space volume fraction	0.2
Minimum pressure	2.32 MPa
Maximum pressure	10.77 MPa
Mean pressure	5.00 MPa
Pressure ratio	4.64
Cold volume/total volume	0.7082

Heat Transferred per Cycle			
Space	In	Out	Net
High temperature	2917 J	2051 J	865 J In
Intermediate temperature	1406 J	4371 J	2965 J Out
Low temperature	7079 J	4979 J	2100 J In

1 Joule = 0.000948 Btu; 1 kW = 3413 Btu/h; 1 cm^3 = 0.061 in^3; 1 MPa = 145 psia.

TABLE 7.16 Ideal Mechanical-Compression Vuilleumier Heat Pump (with 40 percent Dead Space) Analysis Results

(Balanced-Compounded Vuilleumier Embodiment)

Cooling capacity @ 1000 rpm	35 kW
Heating capacity	49.5 kW
Total working volume	4353 cm^3
Dead-space volume fraction	0.4
Minimum pressure	2.95 MPa
Maximum pressure	8.48 MPa
Mean pressure	5.00 MPa
Pressure ratio	2.87
Cold volume/total volume	0.7082

Heat Transferred per Cycle			
Space	In	Out	Net
High temperature	4100 J	3235 J	865 J In
Intermediate temperature	2470 J	5435 J	2965 J Out
Low temperature	9950 J	7850 J	2100 J In

1 Joule = 0.000948 Btu; 1 kW = 3413 Btu/h; 1 cm^3 = 0.061 in^3; 1 MPa = 145 psia.

TABLE 7.17 Ideal Ericsson-Ericsson Heat Pump (with 20 percent Dead Space) Analysis Results

(Ericsson-Ericsson Embodiment)

Cooling capacity @ 1000 rpm	35 kW
Heating capacity	49.5 kW
Total working volume	3340 cm^3
Dead-space volume fraction	0.2
Minimum pressure	3.43 MPa
Maximum pressure	7.28 MPa
Mean pressure	5.00 MPa
Pressure ratio	2.12
Cold volume/hot volume	2.4274
Interpiston volume/total volume	0.62

Heat Transferred per Cycle			
Space	In	Out	Net
High temperature	1,948 J	1,083 J	865 J In
Intermediate temperature	8,132 J	11,097 J	2,965 J Out
Low temperature	4,728 J	2,628 J	2,100 J In

1 Joule = 0.000948 Btu; 1 kW = 3413 Btu/h; 1 cm^3 = 0.061 in^3; 1 MPa = 145 psia.

TABLE 7.18 Ideal Ericsson-Ericsson Heat Pump (with 40 percent Dead Space) Analysis Results

(Ericsson-Ericsson Embodiment)

Cooling capacity @ 1000 rpm	35 kW
Heating capacity	49.5 kW
Total working volume	6029 cm^3
Dead-space volume fraction	0.4
Minimum pressure	3.79 MPa
Maximum pressure	6.59 MPa
Mean pressure	5.00 MPa
Pressure ratio	1.74
Cold volume/hot volume	2.4274
Interpiston volume/total volume	0.62

Heat Transferred per Cycle			
Space	In	Out	Net
High temperature	2,463 J	1,598 J	865 J In
Intermediate temperature	11,595 J	14,560 J	2,965 J Out
Low temperature	5,979 J	3,879 J	2,100 J In

1 Joule = 0.000948 Btu; 1 kW = 3413 Btu/h; 1 cm^3 = 0.061 in^3; 1 MPa = 145 psia.

would carry its heat from the warmer to the cooler spaces and substantially reduce the thermodynamic performance of the heat pump. The performance of equipment without regenerators is so poor as to be impractical. Much theoretical and experimental work has been devoted to improving regenerator design.

A 100% efficient regenerator captures and stores all the heat from the working fluid as it passes to the cooler space and delivers this heat back to the working fluid when it returns to the warmer space. By this operation, the regenerator blocks the heat flow between spaces. Regenerators with less than 100% efficiency allow some heat to pass between the spaces.

Nonideal regenerator effects depend on where they occur in the cycle. In the refrigerator segment, the heat flow past the regenerator into the cold space heats the cold space and reduces its capacity to absorb heat from the environment. For the same capacity, the size of the heat pump must be increased to offset this inefficiency. This heat recirculates inside the refrigerator segment and does not affect the net heat exchanged with the environment.

In the engine segment, the heat flow past the regenerator is in the same direction as the heat flow due to the cycle operation. It simply adds to the heat absorbed and rejected by the working fluid at the high-temperature and intermediate-temperature levels, respectively. Additional heat input is required to maintain the high temperature at the hot space; otherwise the heat flow will begin to cool the hot space. Additional heat rejection is also required to prevent the warm-space temperature from rising. This heat flow increases the heat delivered by the heat pump at the intermediate temperature, but does not add to the required working volume.

Evaluation of nonideal regenerator effects requires determination of the maximum and minimum amounts of working fluid in each space. These extremes do not necessarily occur when the volumes are at their maximum and minimum. For example, a smaller volume at a higher pressure could result in more working fluid at that temperature than at the maximum working volume. Once these fluid limits are known, the difference is the amount of fluid that went through the regenerator to a space associated with a different temperature. The amount of heat that must be stored by the regenerator is the product of the heat capacity of the fluid, the amount of fluid traveling through the regenerator, and the temperature difference between the two spaces on either side of the regenerator.

To find the amount of fluid transferred, the hot- and cold-space analyses were modified to report the maximum and minimum fluid volumes. For Stirling-Stirling machines, this is the same as calculating the fluid volumes for the warm spaces. However, Vuilleumier machines have a combined warm space, and knowing the maximum and minimum fluid amounts does not indicate whether the fluid is traveling to the hot space or the cold space.

For our calculations, thermodynamic properties for helium were used. Helium is the working fluid usually selected for these systems. It has high heat transfer and low fluid friction loss characteristics, similar to hydrogen, but is considered safer than hydrogen. Other potential working fluids have lower heat-transfer coefficients and higher fluid friction losses and are not often used in these systems. However, using air as a working fluid can have practical advantages that outweigh these considerations, such as that the equipment can be made self-recharging to compensate for leaks, because it is immersed in a working-fluid supply.

Table 7.19 shows the amount of energy that must be stored in a 100% efficient regenerator. The impact of nonideal effects on a system is dependent on the amount of energy the regenerator has to cyclically store and release. As seen in Table 7.1, the thermal-compression Vuilleumier embodiment absorbs 2100 J (1.99 Btu) per cycle at the low temperature. As shown in Table 7.19, for 20% dead space volume, the regenerator between the low and intermediate temperatures must cyclically accept and deliver 5149 J (4.88 Btu) per cycle. If this regenerator is 90 percent efficient, 10 percent of this heat is instead carried from the warm to the cold space. This causes the refrigeration segment to carry the 2100 J (1.99 Btu) for the cooling effect from the cold

TABLE 7.19 Comparison of Regenerator Heat Storage Requirements

System	No dead space		20 percent dead space		40 percent dead space	
	Hot-warm, J/cycle	Warm-cold, J/cycle	Hot-warm, J/cycle	Warm-cold, J/cycle	Hot-warm, J/cycle	Warm-cold, J/cycle
Thermal-compression Vuilleumier (traditional Vuilleumier & Vuilleumier with internal heat exchangers)	16,330	4,109	20,329	5,149	27,041	6,883
Piston-piston Stirling-Stirling (balanced-compounded Vuilleumier)	1,023	971	1,321	1,293	1,816	1,799
Displacer-piston Stirling-Stirling (duplex Stirling)	2,217	2,018	3,141	2,644	4,568	3,265
Mechanical-compression Vuilleumier (balanced-compounded Vuilleumier)	1,078	1,078	1,347	1,347	1,831	1,831
Ericsson-Ericsson	1,890	1,890	2,390	2,390	3,217	3,217

1 Joule = 0.000948 Btu.

to the warm space, only to have 515 J (0.49 Btu) of heat return to the cold space because of the inefficiency of the regenerator. Because 515 J (0.49 Btu) must be moved to the warm side (only to flow back to the cold side), the ability to absorb heat from the environment is reduced by 515 J (0.49 Btu) to only 1585 J (1.50 Btu). To accommodate the additional heat load and maintain a net heat absorption of 2100 J (1.99 Btu) from the environment, the system's size must be increased. However, the heat loss from the warm to the cold space is not constant, but changes with the working volume. More precisely, this heat loss is a function of the amount of fluid present, with the loss increasing as the amount of fluid increases. The required working volume was sized by iteratively calculating the heat flow through the regenerator between the spaces, then calculating the working volume required to lift the absorbed heat along with the heat returned internally through the regenerator, until convergence is achieved. The new working volume needed to accommodate the heat flow through the cold-space regenerator is 6826 cm^3 (416.5 in^3), 32 percent larger than that found by the ideal analysis. This volume increase requires a net heat input increase at the high temperature of the ratio of the new volume to the ideal volume; that is, the original 865 J (0.82 Btu) per cycle increases to 1146 J (1.09 Btu).

Similarly, the volume increase implies that more heat must be stored in the engine regenerator, increasing it from 20,329 J to 26,933 J. The regenerator between the hot- and warm-space volumes was also assumed to be only 90 percent efficient. Therefore, 2693 J (2.55 Btu) flows from the hot to the warm space per cycle (this value also increases with increased working volume). During operation of the thermal-compression Vuilleumier embodiment, this energy loss must be made up by additional heat input to the hot space.

The net effect is to increase the required heat input from 865 J (0.82 Btu) to 1146 J + 2693 J = 3839 J (3.64 Btu) per cycle. An alternative way to look at this effect is that the cooling COP (coefficient of performance) has been reduced from 2.43 (2100 J/865 J) to 0.55 (2100 J/3839 J).

If heating is the desired heat pump output, the effect of regenerator inefficiencies on the COP is greatly reduced. The heat rejected is the sum of the heat energy absorbed at the high and low temperatures. The heating COP of the thermal-compression Vuilleumier embodiment with 100 percent effective regenerators is 3.43 [(2100 J + 865 J)/ 865 J]. The heating COP for this Vuilleumier embodiment with 90 percent effective regenerators is 1.55 [(2100 J + 3839 J)/3839 J]. In heating, the additional heat flow from the high- to the intermediate-temperature side of the regenerator contributes to the useful output. However in cooling, the additional heat flow only adds to the required heat input, not to the useful cooling effect.

For the mechanical-compression Vuilleumier, the new working volume required to meet the design capacity and regenerator inefficiency is 2461 cm^3 (150.1 in^3), or 7 percent more volume. The intermediate- to low-temperature net heat flow through the refrigerator regenerator is 144 J (0.14 Btu). The increase in heat input due to the increase in volume is from 865 J (0.82 Btu) to 924 J (0.88 Btu) per cycle. The net heat flow from the high to the intermediate temperature through the engine regenerator is 144 J (0.14 Btu), which must also be added to the heat input for the balanced-compounded Vuilleumier to operate. The resulting heat input per cycle is 924 J + 1445 J = 1068 J (1.01 Btu), up from 865 J (0.82 Btu).

The cooling COP, accounting for regenerator inefficiencies is reduced from 2.43 (2100 J/865 J) to 1.97 (2100 J/1068 J). The heating COP is reduced from 3.43 [(2100 J + 865 J)/865 J] to 2.97 [(2100 J + 1068 J)/1068 J].

Heat pumps with smaller working-fluid volumes are less affected by nonideal regenerator effects. From the above calculation the 10% inefficiency causes the thermal-compression Vuilleumier embodiment to drop in cooling COP from 2.43 to 0.55, whereas the cooling COP of the mechanical-compression Vuilleumier embodiment drops only from 2.43 to 1.97.

Regenerators with high efficiency are very beneficial to integrated heat pumps. Any inefficiency in the warm-to-cold regenerator must be compensated for by an increase in working volume (or by other means). This volume increase puts more load on the regenerator. Any inefficiency in the hot-to-warm regenerator must be compensated for by additional heat input at the high-temperature end of the regenerator. The combined effect is to reduce both the heating and cooling COPs. Although the first Stirling engines operated with no regenerator element, the goal of those machines was to produce useful work, without any emphasis on efficiency.

Heat pump embodiments with a low regenerator storage requirement are inherently less sensitive to regenerator efficiency. Practically, the volume of the regenerator acts as dead space for the cycle. A regenerator required to store 500 J/cycle (0.47 Btu/cycle) will add less dead space than a regenerator that stores 5000 J/cycle (4.74 Btu/cycle). The inefficiency of a regenerator is in part due to heat transfer considerations, and having a temperature difference that causes the heat to flow between the working fluid and the regenerator. Although a 100 percent efficient regenerator cannot be practically achieved, striving for the maximum achievable efficiency should be a high priority in designing an integrated heat pump.

Varying heat flux for isothermal operation

In continuous-flow heat pumps operating at steady-state, the heat transfer rates in the heat exchangers are constant. However, the heat pump embodiments in this book operate cyclically, absorbing heat into a space during part of the cycle and rejecting heat from another space during the other portion of the cycle. Because of this periodic heat flow, heat exchangers need to be sized to accommodate the higher heat transfer rate in a given step rather than the lower net average rate.

In an ideal Stirling engine, heat is transferred to the fluid while it is in the hot space during the expansion step (hot piston moving, warm piston stopped). The fluid is then moved to the warm space (through the regenerator), where heat is transferred out during the compression step (hot piston stopped, warm piston moving). The fluid is then returned to the hot space. Because of the discontinuous motion of the pistons, the fluid resides in the space where heat transfer is occurring.

In more realistic (nonideal) cycle embodiments, where the piston (or displacer) motion is continuous, the working fluid is present in two or more spaces at the same time. To maintain isothermal conditions in all spaces, heat must be exchanged with the working fluid. The heat exchanged depends on both the amount of fluid in the space and the pressure changes (which try to shift the temperatures) produced by the cycle.

To determine the amount of cyclic heat transfer that is required to maintain isothermal conditions in the expansion and compression spaces, the heat flows into and out of the spaces were added. Tables 7.1 to 7.6 and 7.9 to 7.18 show the heat flow in and out, in addition to the net heat flows for each of the spaces. The heat-flow amounts, excluding the net heat flow, are those required for isothermal operation. These values are comparable among the concept embodiments, because each embodiment is sized for the same capacity and all have been analyzed under the same assumptions.

One way to visualize how the cyclic heat transfer keeps the working fluid isothermal is to envision the gas as interacting with an external heat reservoir. In a real embodiment, the medium interacting with the working fluid, along with the thermal mass of the elements containing the working fluid (including the heat exchangers), acts as this reservoir. For example, in a real machine, when the gas warms up, it loses heat to the piston and cylinder, warming them up and decreasing the gas temperature. When the machine cools, the gas picks up some heat from the piston and cylinder, again moderating the temperature swing. If the heat capacity of the piston and cylinder is large enough and the heat transfer rate between wall and gas is high enough, the gas is kept at a constant temperature. If either the heat capacity or the heat transfer rate is too low, the compression and expansion are closer to adiabatic than isothermal, and the refrigeration capacity of the system is reduced. Heat is then carried to the heat exchangers and regenerators.

The cyclic heat transfer values are an approximate measure of how easy it is to maintain isothermal operation. The larger the value, the more difficult it will be to keep the gas isothermal.

Comparing Tables 7.1 through 7.5 shows that the balanced-compounded Stirling (piston-piston Stirling-Stirling embodiment) is the easiest to keep isothermal. For isothermal operation, the balanced-compounded Stirling engine with no dead space requires only 32 percent more heat cyclically transferred into the hot space than the net heat absorbed into the working fluid. Also, the engine warm space rejects over twice as much heat per cycle as the hot space.

With the addition of dead space, the heat required to maintain isothermal conditions increases. This is because more working fluid (required to achieve the desired mean operating pressure when the volume is increased) must be kept isothermal. Comparison of Table 7.3 with Table 7.8 shows that to maintain isothermal conditions for the larger gas volume needed for 20 percent dead space, the hot space of the balanced-compounded Stirling engine actually needs to absorb 56 percent more heat than is utilized in the cycle. Also, the engine warm space absorbs almost twice as much heat as the net heat rejection.

The piston-piston Stirling refrigerator also needs to transfer more heat through its heat exchangers than the net amount transferred. However, the total heat transferred is less than for the engine because the temperatures of the spaces are closer.

Table 7.4 shows the heat transfer rates for the displacer-piston Stirling embodiment (duplex Stirling). This Stirling-Stirling configuration, which has a slightly lower pressure ratio and larger working volume than the piston-piston Stirling-Stirling, also has more heat cycled to maintain isothermal conditions.

Table 7.5, which contains the mechanical-compression Vuilleumier embodiment (balanced-compounded Vuilleumier) results, shows the heat transfer rates for a heat pump with more working fluid in the system. The heat transferred in at the hot space is 2.4 times the net amount of heat influx.

As shown in Table 7.1, the thermal-compression Vuilleumier embodiments, which have the largest working volume and the lowest specific capacity, have the largest cyclic heat transfer values. Thus, compared with the other embodiments, thermal-compression Vuilleumier heat pumps are the most difficult to maintain in isothermal operation.

Table 7.6 shows that the Ericsson-Ericsson heat pump has a relatively small value for cyclic heat transfer at the high and low temperatures, while it has a very large value at the intermediate temperature. This is a result of the large fraction of the total volume devoted to the space between the flywheel pistons.

In a real heat pump isothermal heat transfer is not achieved. Heat transfer between the working fluid and the environment occurs only with temperature differences. For heat to flow into the gas, the gas temperature must be lower than the heat-exchanger wall temperature, which must be lower than the environment (air, water, combustion gases) temperature. The reverse must be true for the gas to reject heat. During the cycle, the temperature differences can decrease, reducing heat flow. The gas-temperature variations are not normally large enough to cause a substantial amount of heat to be rejected where it is normally absorbed, and absorbed where it is normally rejected.

Mechanical energy storage

In addition to storing heat, the mechanical-compression concept embodiments need to store and retrieve work during their cycles. The thermal-compression Vuilleumier embodiments do not deliver or absorb work because the working fluid distributes itself so that the power generated by the engine segment is used by the refrigerator segment at the same instant. The movement of the fluid between the engine and refrigerator segments and their pressure changes are the work transfer mechanisms.

The mechanical-compression systems operate differently because their working volumes are changing, and p-V work is created and utilized by the engine and refrigerator segments, respectively, at different times during the cycle. In heat pump cycles the amount of work delivered back to the process to compress the working fluid differs from that generated when the working fluid was expanding. The net work generated by the engine segment (the difference between what was generated during the expansion of the fluid and what was consumed during compression) is absorbed by the refrigerator segment. However, this does not necessarily happen in such a way that the work is generated at the instant it is needed. To balance the work flow, the work needs to be temporarily stored. The concept embodiments presented in this book temporarily store energy in the momentum of the moving elements. The piston in the duplex Stirling, the pistons in the balanced-compounded Stirling and balanced-compounded Vuilleumier, and the flywheel pistons of the Ericsson-Ericsson are designed to be energy-storing elements. As the pistons accelerate, the work is stored in their increased momentum. The work is delivered back to the working fluid as the pistons slow down.

In addition to work storage, the mass of the pistons permits the proper phasing of the volume changes. The displacers and pistons must move out of phase with each other, so that expansion of the working fluid occurs when the fluid is in the hot or cold space, and compression occurs when it is in the warm space. Examining one cycle of the balanced-compounded Stirling exemplifies the requirement for phasing the motion of the moving elements. If they both compress and expand the working fluid at the same time (180° out of phase), there will be no transport of fluid between the different temperature spaces to achieve the required heat transfer. If they move in phase with each other, so that one volume increases as the other decreases, then the working fluid will be transferred between the two spaces and the heat transfer can occur. Although the pressure will vary as the temperature is changed, no external work will be generated or consumed because no volume change has occurred. By phasing the pistons by 90° (for details, see the cycle operation description in Chap. 5), the required pressure and volume changes are produced to generate work (for the engine) or absorb work and move heat to a higher temperature (for the refrigerator).

In four-cycle equipment, the pistons' operation is phased by 90° (one-quarter of a cycle). In single-cycle equipment, the relatively large inertia of the pistons causes them to respond slowly to fluid forces on them. This is unlike the displacers, which are lightweight and react quickly to the fluid forces. This difference in reaction times creates the phasing required for the volume changes.

The work-storage requirement calculations were done assuming sinusoidal motion of all elements. The calculation averages the sums of the p-V work for each space in and out (taking the average of the sums reduces any numerical inaccuracies that can occur). The results of these calculations are shown in Table 7.20. The thermal-compression Vuilleumier embodiments do not alter their working volumes, so no work is stored during the cycle. The rest of the concept embodiments are ordered by increasing storage requirements.

The results presented are for an ideal cycle. If the system is built as a free-element embodiment (not constrained by a crank), the components of a real embodiment will deviate from the ideal motion used in the analysis. Any mechanism that controls the motion of the elements will eliminate this deviation.

Each concept embodiment provides its own method of energy storage. If any of these were driven by a crank mechanism, storage would be provided by the mass of the pistons and the crank, and would probably include a flywheel. However, as these concepts have been configured, they use free elements. The duplex Stirling embodiment of the displacer-piston Stirling-Stirling concept temporarily stores the excess work in the inertia of the relatively heavy piston. The balanced-compounded embodiments of the piston-piston Stirling-Stirling and the mechanical-compression Vuilleumier use the pistons as storage elements. The Ericsson-Ericsson uses the flywheel pistons for storage.

TABLE 7.20 Comparison of Work-Storage Requirements

System	Work storage, J/cycle		
	No dead space	20 percent dead space	40 percent dead space
Thermal-compression Vuilleumier (traditional Vuilleumier & Vuilleumier with internal heat exchangers)	0	0	0
Piston-piston Stirling-Stirling (balanced-compounded Stirling modeled as a single cycle)	3,280	4,816	7,120
Displacer-piston Stirling-Stirling (duplex Stirling)	3,711	5,330	7,930
Mechanical-compression Vuilleumier (balanced-compounded Vuilleumier modeled as a single-cycle)	5,687	8,719	12,992
Ericsson-Ericsson	6,251	8,091	11,073
Balanced-compounded Vuilleumier (four-cycle mechanical-compression Vuilleumier)	669	not calculated	not calculated

1 Joule = 0.000948 Btu.

As a general rule, the more energy that is stored, the larger the mass required to contain it. For the same operating speed the piston velocities will be similar. Since the energy stored is a function of velocity, the larger storage requirements will need larger masses. In real equipment, the displacer and piston motion will not be sinusoidal and will vary, along with the phasing, as the components react differently to the fluid forces. However, other investigators[1] have previously concluded that sinusoidal motion is a reasonable approximation, especially for this type of second-order comparative analysis. Thus, the relative importance of the non-sinusoidal motion should not change among the concept embodiments.

The analysis described above and the results in the tables consider only single-cycle concept embodiments. However, the four-cycle balanced-compounded embodiments offer some benefit. With four cycles of equal size, each is separated from its neighbors by a 90° phase angle, so that each cycle can drive the other cycles during part of its operation, reducing the work-storage requirements. To assess this benefit, the work in and out in 1° steps for the balanced-compounded Vuilleumier was divided by 4 (to divide it among the four cycles). The second, third, and fourth cycles were then phased by 90°, 180°, and 270°, respectively. The algebraic sum of the work in and out along each step over the four cycles represents the net amount which must be stored or delivered. The first effect seen was that the four segments produce four distinct work storage and withdrawal cycles during one complete thermodynamic cycle. The amount of work stored is approx-

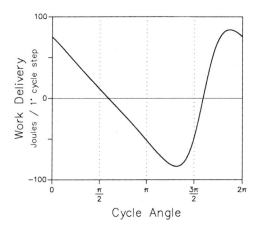

Figure 7.4 Work delivery during one cycle to the moving elements of a mechanical-compression Vuilleumier (35-kW cooling capacity, 1000 rpm).

imately one-fourth the single-cycle variation. In general, this work-storage reduction makes it easier to design and operate a real system.

Figures 7.4 and 7.5 show the work delivery rates for a single-cycle mechanical-compression Vuilleumier embodiment and for a four-cycle balanced-compounded Vuilleumier with the same capacity, respectively. These figures show that four-cycle balanced compounding smooths and reshapes the profile. The single-cycle mechanical-compression Vuilleumier embodiment (one cycle of the balanced-compounded Vuilleumier) profile has a definite skew to it; the storage and retrieval of work have exactly opposite patterns, one rate rising slowly and dropping off quickly, the other rising quickly and dropping off gradually. When this profile is split, rotated, and combined as described above, the work flow appears much more uniform and periodic. In Fig. 7.5 a sine curve of the same amplitude, phased to match the

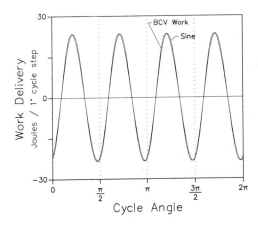

Figure 7.5 Work delivery during one cycle to the moving elements of a balanced-compounded (mechanical-compression) Vuilleumier (35-kW cooling capacity, 1000 rpm).

work flow, has been superimposed on the work flow. This illustrates how closely the work flow in the balanced-compounded embodiments follows a sine curve.

Summary

In general, no concept embodiment has an overwhelming rationale for being selected over another for the selected range of operating conditions. The analysis has shown that embodiments of all the concepts are feasible, but not necessarily for the same type of duty. All have the same ideal heating and cooling capacities and respective COPs. They require similar materials and construction techniques and similar design and manufacturing technology expertise. Heat exchangers and regenerators are common elements, although different sizes are required. However, our results permit some differentiation among the concepts, with some concepts preferred in certain capacity ranges.

Thermal-compression concepts

The thermal-compression Vuilleumier embodiments have inherently larger working volumes than any of the mechanical-compression concepts, but their mechanization is simpler. Larger size is not always a disadvantage. A given clearance tolerance between a displacer or piston and the cylinder wall represents a relatively larger gap for working-fluid leakage in a 1-in-diameter cylinder than in a 3-in-diameter cylinder. If clearance tolerances are critical, larger parts could cost less to mass-produce and assemble. Therefore thermal-compression Vuilleumier concepts are more adaptable to smaller capacities.

Thermal-compression Vuilleumier performance is also more tolerant of larger dead-space volume than the mechanical-compression concepts. Although their working volume for a given capacity is higher, the thermal-compression Vuilleumier embodiments add proportionally less total volume to accommodate the same volume of dead space while maintaining the design capacity. Because the dead-space volume fraction is a larger absolute volume than for mechanical-compression concepts, the heat-exchanger and regenerator volumes (which constitute most of the dead space) can be made larger to improve their performance.

The relatively low specific capacity of thermal-compression Vuilleumier embodiments works against them in larger-capacity units. Since they require more working volume than mechanical-compression embodiments, they may cost more because of the greater amount of material needed to construct them. In addition, the lower capacities and resulting larger working-fluid quantities mean that more is demanded of the heat exchanger and regenerators. This is evident from a comparison of the amount of energy cyclically stored in the regenerators. The thermal-compression Vuilleumier regenerator requires about an order of magnitude more storage capacity than mechanical-compression concept regenerators to store 100 percent of the energy in the working fluid. Thus, these larger capacities could easily demand so much of the regenerator that either (1) the pressure drop through it, (2) the energy stored by it, or (3) its dead space decreases the performance to the point that the heat pump performance is no longer acceptable.

Mechanical-compression concepts

The need for higher capacity provides the rationale for use of mechanical-compression concepts. Their larger heat pump capacity within a reasonable overall volume gives them an advantage over the thermal-compression Vuilleumier embodiments. The smaller demands they place on their heat exchangers and regenerators mean that moderate inefficiencies in heat transfer degrade their performance less than that of the thermal-compression Vuilleumier embodiments. However, the mechanical-compression concepts are clearly in two separate groups: four-cycle and single-cycle machines. These groups also show performance distinctions based on their size. The balanced-compounded embodiments, composed of four cycles, distribute their capacity among four sets of heat exchangers and regenerators. Tooling and assembling very small heat exchangers and moving parts can be difficult. Therefore, for smaller-capacity equipment, it should be easier to make the duplex Stirling or Ericsson-Ericsson heat pumps than either of the balanced-compounded embodiments. However, for larger-capacity equipment, the balanced-compounded embodiments offer the benefit of having a smaller work-storage requirement than the single-cycle embodiments.

Heat exchangers and regenerators also limit the size of the mechanical-compression embodiments. Using a multiple-cycle embodiment extends the applicable range of capacities. Also, larger components usually operate at slower speeds and require a larger working volume to compensate. Attempting to speed up the reciprocating elements can cause unacceptably high stress levels on the mechanical linkages. Vibration caused by reciprocating motion cannot be totally eliminated. This would also be true of the duplex Stirling and Ericsson-Ericsson. However, the balanced-compounded embodiments couple multiple units together and phase their operation to reduce the vibration.

TABLE 7.21 Summary of General Observations on Heat Pump Concepts

\multicolumn{2}{c}{Class: Thermal compression (thermal-compression Vuilleumier, with external or internal heat exchangers)}	
Size range	Lower end of capacity range
Assets	Lower specific capacity can make small-capacity machines manageable
	Less sensitivity to dead space (partly a result of starting with a larger volume) suggests that incorporating real heat exchangers and regenerators may be easier than with other concepts
Liabilities	Large capacities not as feasible
	Large working-fluid volumes make concepts more sensitive to heat-exchanger and regenerator inefficiencies
	Requires an external driving force to move the displacers
\multicolumn{2}{c}{Class: Single-cycle mechanical compression (duplex Stirling, Ericsson-Ericsson)}	
Size range	Middle range, overlapping with the other concepts, but probably centered around residential capacity, 10 kW cooling (3 tons)
Assets	Small size for given capacity
	Lower sensitivity to regenerator and heat-exchanger efficiency
	Self-starting/self-running (according to inventors)
	Fewer heat exchangers and regenerators than the four-cycle concepts
Liabilities	Requires close tolerances on the displacer and piston masses to match the operating characteristics and achieve phasing
	Capacity is limited by the mass of the pistons, which must not be too heavy to be supported or driven. Tendency for too much undamped vibration
	Single-cycle operation requires some balancing mechanism to dampen vibration; the Ericsson-Ericsson provides one mechanism by using counteracting flywheel pistons
\multicolumn{2}{c}{Class: Multiple-cycle mechanical compression (balanced-compounded Stirling, balanced-compounded Vuilleumier)}	
Size range	Medium and larger capacities (overlapping the single-cycle range); smallest size probably limited by manufacturing considerations
Assets	Small size for given capacity (relatively large specific capacity)
	Lower sensitivity to regenerator and heat-exchanger inefficiency
	Self-starting/self-running (according to inventor)
	For the balanced-compounded Stirling, concept provides balanced motion of elements to reduce vibration during operation (the balanced-compounded Vuilleumier provides some balance, but not as complete as the Stirling)
	Uses the same replicated components and does not rely on the matching of light and heavy masses
	Four-cycle configuration allows one cycle to deliver work to others, reducing the dependency on moving masses for work storage
	Use of four cycles provides automatic control over phasing of piston motion
Liabilities	Potentially limited for low-capacity designs because of limitations of manufacturing technology

Free-piston configuration considerations

Heat pumps which use the inertia of piston elements for cyclic work storage in their operation cannot be scaled up indefinitely. Even though the duplex Stirling and Ericsson-Ericsson embodiments use no mechanical linkages, they rely on masses which must be moved by the working fluid. One potential limiting factor is that the high mass of the pistons can lead to unacceptable friction losses. This is why small duplex Stirling heat pumps are usually configured vertically. This is also why magnetic and gas bearings are considered. Another potential limitation is the size of the displacers that can be manufactured with a sufficiently low weight so that they can quickly respond to gas pressures while maintaining structural integrity. Although a rigorous analysis of the dynamics coupled with experiments is needed to identify these limits, the primary researchers of these concepts believe that embodiments can be made with capacities into the light commercial range of cooling capacities, 35 kW (10 tons).

The balanced-compounded embodiments have a higher capacity ceiling than the other mechanical-compression concept embodiments. Because their capacity requirements are distributed over four distinct cycles, their operation depends less on piston inertia and less on the mechanical design. The four cycles automatically operate with a stable 90° phase angle. Also, their piston assemblies have approximately equal weight, avoiding the more complex balancing of the lightweight displacers and heavy pistons to control the phasing. The balanced-compounded embodiments also have a higher-capacity design limit because their cyclic work-storage requirements are lower, and lighter piston assemblies can be used for the same capacity.

Table 7.21 summarizes the analysis results for thermal-compression, single-cycle mechanical-compression, and multiple-cycle mechanical-compression heat pumps.

Reference

1. Vincent, R. J., W. D. Rifkin, and G. M. Benson, "Analysis and Design of Free Piston Stirling Engines—Thermodynamics and Dynamics," paper 809334 presented at the 15th Intersociety Energy Conversion Engineering Conference, Seattle, Aug. 18–22, 1980.

Chapter

8

Analysis of Real Integrated Heat Pumps

Several independent developments of integrated heat pumps have been pursued in the United States, Germany, and Japan. Although none have yet reached commercial status, they show a great deal of promise. The first half of this chapter describes the equipment that has been developed and its current status.

The second half of the chapter describes design optimization calculations that remove some of the assumptions of the comparative analysis presented in Chaps. 6 and 7. This optimization accounts for the effects of using realistic heat exchangers, which have dead space and flow friction losses and do not provide perfect heat exchange. As examples, two embodiments are analyzed: a thermal-compression Vuilleumier heat pump and a Stirling-Stirling heat pump.

Recent Hardware Development Programs

During the past 20 years, research and development has brought some integrated heat pump concepts close to commercialization. This section describes several of these efforts.

Philips Research Laboratories

Philips has been a modern pioneer in Stirling engine and refrigerator technology. Its research on one- and two-stage Stirling-cycle machines for cryogenic applications began in the 1950s and led to small-capacity Stirling refrigerators of approximately 1 W (3.4 Btu/h) capacity, which are used for cooling infrared detectors.[1,2,3]

In the early 1960s, Philips began investigating the use of Vuilleumier-cycle machines for cryogenic applications. This research and development continued through the 1970s, culminating in the development of several packaged prototypes. However, these Vuilleumier prototypes have not yet led to commercial products.

In the 1980s Philips oriented its heat pump research and development towards space heating. Figure 8.1 is a drawing of a 10-kW (34,000-Btu/h) Vuilleumier heat pump, and Fig. 8.2 shows the housing of the experimental prototype. The cylinder on the right contains the electric drive motor. The refrigerator cylinder is on the left, and the engine cylinder is at the top. This unit was electrically heated, presumably for better experimental control and measurement of heat inputs. In the late 1980s, Philips stopped working on Stirling- and Vuilleumier-cycle refrigerators.

Professor Eder's Vuilleumier heat pumps

Professor Franz Xaver Eder's work at the Technical University of Munich in West Germany has led to significant advances in small-capacity integrated heat pumps. Professor Eder, a cryophysicist, recognized the potential benefits of integrated heat pump cycles for space conditioning and power generation in the late 1970s. In 1981 Profes-

Figure 8.1 Vuilleumier heat pump prototype layout. (*Courtesy of Philips Research Laboratories.*)

Figure 8.2 Vuilleumier heat pump prototype. (*Courtesy of Philips Research Laboratories.*)

sor Eder reported on his integrated heat pump research[4] at the International Institute of Refrigeration Joint Meeting of Commissions B1, B2, E1, and E2. Since then, Professor Eder and his colleagues at the Technical University of Munich have produced laboratory prototypes of integrated heat pumps for residential space heating in cooperation with several West German companies: ASK-Technische Entwicklungen of Bayreuth, PETRY of Neumarkt, and Heidelberg-Motor GmbH of Starnberg.

Figure 8.3 shows one of Professor Eder's early laboratory prototypes, built by PETRY. The unit was intended to produce hot water for hydronic heating, which is the most common space heating method in Europe. The water tank is at the left rear in the photograph. The heat pump is on the table at the center, with the hot cylinder extending vertically upward. This prototype was electrically heated for experimental convenience. Table 8.1 lists the design specifications for this traditional Vuilleumier concept adapted for residential space heating and water heating. Figures 8.4 and 8.5 illustrate a more advanced gas-fired experimental prototype. For these prototypes to be converted into commercial products, they would have to be simplified to achieve the necessary reliability and ease of manufacture. Professor Eder's recent machines with more compact configurations represent significant

Figure 8.3 Vuilleumier heat pump prototype. (*Courtesy of Prof. F. X. Eder and the Technical University of Munich.*)

TABLE 8.1 Nominal Design Specifications for the Vuilleumier Heat Pump Prototype Shown in Fig. 8.3

Working fluid	Helium
Mean operating pressure	6 MPa
Heat input	3 kW
Cooling capacity	7 kW
Driving power input	1 kW
Total heat output	10 kW
Heater temperature	600°C (1100°F)
Delivered temperature	50°C (120°F)
Source temperature	6°C (43°F)
Operating speed	11.1 Hz
Piston stroke	6 cm
Cold cylinder volume	796 cm^3
Hot cylinder volume	623 cm^3

1 kW = 3413 Btu/h, 1 cm^3 = 0.0610 in.^{3}, 1 MPa = 145 psia.

progress toward a practical design.[5] These machines, which are similar to those of Cooke-Yarborough,[6] use compact heat exchangers together with a simplified driving mechanism. Improvements in design and performance over those of the machines shown in Figs. 8.4 and 8.5 were reported to the Society of German Air Conditioning and Refrigerating Engineers in 1989.[7] Professor Eder estimates that these design improvements make a COP$_h$ greater than 2 achievable for Vuilleumier heat pumps used for space heating.[5]

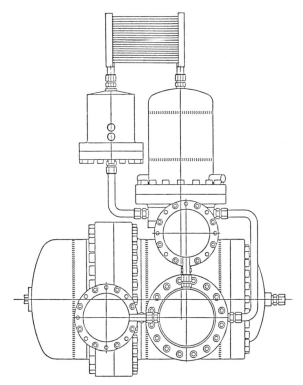

Figure 8.4 Layout of an advanced gas-fired Vuilleumier heat pump prototype. (*Courtesy of Prof. F. X. Eder and the Technical University of Munich.*)

Sunpower's duplex Stirling heat pumps

Sunpower, Inc. of Athens, Ohio, has been developing duplex Stirling heat pumps since 1979. This development builds on the free-piston Stirling engines conceived in the late 1960s by the founder of Sunpower, Professor William Beale.[8] His regenerative free-piston concept became a keystone in the development of many Stirling engine and refrigerator embodiments. The Sunpower duplex Stirling laboratory prototypes include a natural gas liquefier, a residential heat pump, and a residential refrigerator.

Natural gas liquefier Liquefying natural gas requires temperatures far lower than those needed for space conditioning. Figure 8.6 shows the first proof-of-concept version of Sunpower's natural gas liquefier. Figure 8.7 shows a later version, which has greater capacity and operates more stably over a full range of cold-end temperatures, from ambient temperature to its design operating temperature. This device includes a 2.5-kW

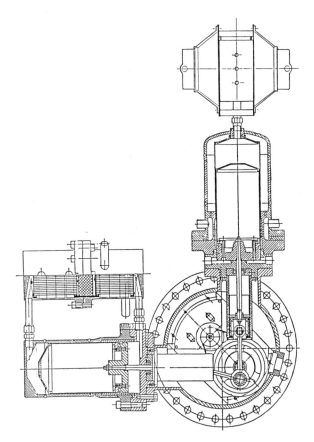

Figure 8.5 Cross section of an advanced gas-fired Vuilleumier heat pump prototype. (*Courtesy of Prof. F. X. Eder and the Technical University of Munich.*)

(8500-Btu/h) heat engine coupled to a refrigerator designed to lift approximately 500 W (1700 Btu/h) from a 110 K (−260°F) heat source.[9] Sunpower's experience in designing, building, and testing these liquefiers was important to its later development of a residential heat pump and a residential refrigerator. The liquefier has not yet been developed far enough to evaluate its commercial potential. However, with the present interest in natural gas-fueled vehicles, small natural gas liquefiers such as this will probably receive more research attention.

Residential heat pump In the early 1980s, Sunpower tested a small residential-size (11 kW, or 36,000 Btu/h) duplex Stirling heat pump. This packaged prototype achieved 50 percent of Carnot COP_h at the low overall temperature lift of 50°C (90°F). More important than the actual performance was Sunpower's demonstration that it was possi-

Analysis of Real Integrated Heat Pumps   131

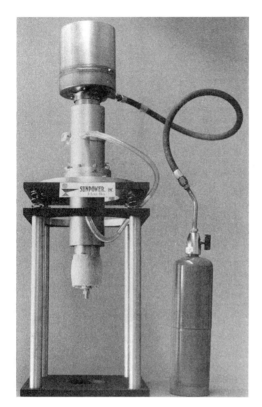

Figure 8.6 Proof-of-concept duplex Stirling natural gas liquefier. (*Courtesy of Sunpower, Inc.*)

ble to design efficient Stirling-cycle heat pumps for the small temperature differences required in a residential application.[10] The only machine built and tested was the one shown in Figs. 8.8 and 8.9. Sunpower has not advanced the design past this packaged prototype, which was developed under Gas Research Institute sponsorship. It did not reach the commercial prototype stage because the equipment costs seemed too high for commercial success at the time.

Residential refrigerator   The free-piston duplex Stirling heat pump has the potential for mechanical simplicity and high reliability in the size range of residential refrigerators, 40 W (140 Btu/h). Figure 8.10 shows a prototype for this application. Table 8.2 gives its specifications.

This mechanically simple free-piston machine can achieve cooling temperatures of about −20°C (−4°F), which are needed for the freezer compartment. At these temperatures, the refrigerator can have adequate cooling capacity to restore the temperature of a food compartment quickly after warm food is placed inside. It runs quietly and in a stable

Figure 8.7 Prototype duplex Stirling natural gas liquefier. (*Courtesy of Sunpower, Inc.*)

manner and is easily controlled. A packaged prototype has demonstrated more than 2000 hours of operation with no measurable wear, showing that it has the potential for long life.[11] Its mechanical simplicity also suggests a potential for low maintenance and ease of manufacture.

University of Dortmund Vuilleumier heat pumps

Several researchers at the University of Dortmund, West Germany, led by Professor S. Schulz, have successfully applied both analytical and experimental approaches to Vuilleumier heat pump development. Like other Vuilleumier-cycle researchers, they have developed models of the performance of Vuilleumier-cycle residential heat pumps.[12] In addition to conventional cycle analysis, they have suggested a model based on discontinuous displacer motion similar to the discontinuous steps used to describe the ideal Stirling and Ericsson cycles in Chap. 3. They have also developed a mathematical expression,[6] based on the doctoral thesis of Norbert Richter,[13] that is useful in optimizing the ratio of the swept volumes of the two cylinders.

A low-speed heat pump (about 7 Hz) with a capacity of 3.5 to 7 kW (12,000 to 24,000 Btu/h) has been built in this continuing program of hardware development. As with all classical Vuilleumier heat pumps, external power is needed to move the displacers. The approach taken at the University of Dortmund incorporates a small Stirling engine cycle into the Vuilleumier mechanism by making the diameter of one displacer shaft larger than that of the other. This adds mechanical compression to the thermal-compression effect of the traditional Vuilleumier cycle. The force differences acting on this displacer generate enough

Analysis of Real Integrated Heat Pumps   133

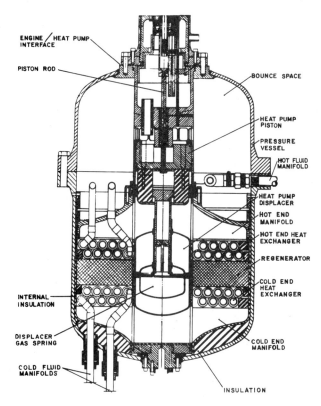

Figure 8.8 Cross section of prototype duplex Stirling residential heat pump. (*Courtesy of Sunpower, Inc.*)

work to drive the displacers. However, the volume change that occurs is very small, so this remains essentially a thermal compression machine. (The mechanical-compression Vuilleumier embodiment described in earlier chapters has much greater volume changes. In those machines, the minimum volume is half the maximum volume, or less.) The small Stirling engine cycle should produce enough power to drive the displacers under most conditions. An electric starter motor will be sealed in the crankcase, because the Stirling-cycle engine power will be available only when the system is running. The motor will also provide power when the output of the Stirling engine cycle is inadequate to drive the displacers. It could also be designed to produce electric power when the cycle produces more work than is needed.

Figure 8.11 shows the general design configuration, and Fig. 8.12 shows the prototype machine. Figure 8.13 shows the moving components, Fig. 8.14 shows the heater head assembly, and Fig. 8.15 shows the burner. The assembled unit is relatively large, typical of a thermal-compression integrated heat pump. It was not designed for

Figure 8.9 Duplex Stirling residential heat pump. (*Courtesy of Sunpower, Inc.*)

small size, but for low speed, ease of instrumenting, and the capability for substituting components of different design. The assembled unit has been used for short-term tests of several heater heads.

Sanyo Vuilleumier heat pump

The idea of more efficient space conditioning with heat-powered heat pumps is very attractive in Japan because that country relies heavily on imported fuels. Use of the heat pumps for cooling will also help reduce peak summer demands for electric power. Sanyo Electric Company, Tokyo Gas Company, Osaka Gas Company, and Toho Gas Company recently announced their joint plan to market a Vuilleumier heat pump[14] in addition to heat pumps driven by electric motors and natural gas engines.

The packaged prototype shown in Figure 8.16 is designed for space heating and cooling and water heating in medium-size shops and offices. The specifications are a cooling capacity of approximately 11 kW (36,000 Btu/h) and a heating capacity of 20 kW (68,000 Btu/h). Sanyo has reported achieving a COP_h of 1.5 and a COP_c of 0.9 through

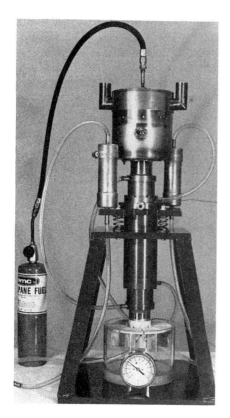

Figure 8.10 Duplex Stirling residential refrigerator (*Courtesy of Sunpower, Inc.*)

TABLE 8.2 Nominal Design Specifications for the Sunpower Residential Refrigerator

Engine power	30–40 W
Heat lifted	40–50 W at −20°C
Working fluid	Helium
Mean operating pressure	1 MPa
Operating speed	32 Hz

1 W = 3.413 Btu/h, 1 MPa = 145 psia.

the use of originally developed simulation software and breadboard development. The prototype is said to be highly controllable, so the heat pump will be able to respond well to load changes. This feature is valuable because a properly designed and controlled variable-capacity heat pump can, over a heating or cooling season, be significantly more efficient than a constant-capacity heat pump operated in on-off fashion. This heat pump should have two other characteristics that are common to Vuilleumier and Stirling machines: quiet, low-emission external combustion and the use of helium instead of halocarbon

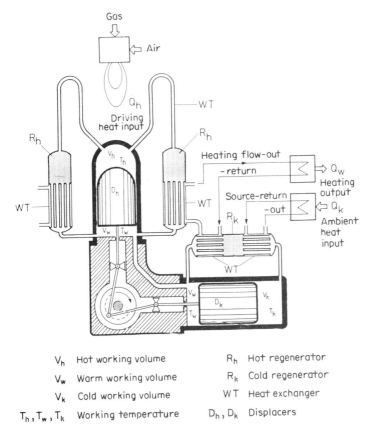

V_h	Hot working volume	R_h	Hot regenerator
V_w	Warm working volume	R_k	Cold regenerator
V_k	Cold working volume	WT	Heat exchanger
T_h, T_w, T_k	Working temperature	D_h, D_k	Displacers

Figure 8.11 Prototype Vuilleumier heat pump configuration. (*Courtesy of Prof. S. Schulz and the University of Dortmund.*)

working fluids. Technical data have not been published, but the product publicity for this development indicates that units will be in use in small- and medium-size stores and offices in the spring of 1992.

Design Optimization

Analysis

Different integrated-cycle heat pumps require very different sizes of heat exchangers and regenerators, even when they operate at the same temperatures, mean cycle pressure, and frequency and have the same ideal cycle capacity. The types and sizes of heat transfer elements play a paramount role in optimal embodiment design. To account for the effects of realistic heat exchangers requires a second-order design technique that includes empirical heat-exchanger design information. The inclusion of this information in a model allows the design to be

Analysis of Real Integrated Heat Pumps   137

Figure 8.12 Prototype Vuilleumier heat pump. (*Courtesy of Prof. S. Schulz and the University of Dortmund.*)

optimized with respect to COP and specific size or capacity. Therefore, we have called this analysis *design optimization analysis*.

Design optimization allows deeper comparative analysis than was presented in Chaps. 6 and 7. The second-order analyses described in those chapters considered two factors that strongly affect heat pump performance: dead space and incomplete regenerative heat exchange. However, those two factors were described parametrically; that is, the performance of the heat pump embodiments was calculated assuming

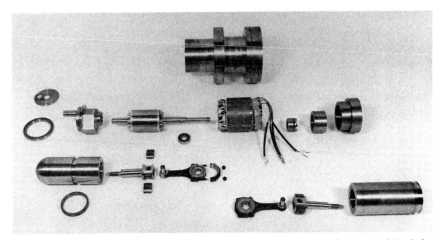

Figure 8.13 Prototype Vuilleumier heat pump moving elements. (*Courtesy of Prof. S. Schulz and the University of Dortmund.*)

Figure 8.14 Prototype Vuilleumier heat pump heater head. (*Courtesy of Prof. S. Schulz and the University of Dortmund.*)

certain fractional dead-space volumes and regenerator efficiencies. In this chapter, we refer to those analyses as parametric. The parametric analyses were based on several other implicit assumptions:

- The heat exchangers were perfect; that is, the temperatures in the working spaces were the same as the temperatures of the external sources and sinks.s

Figure 8.15 Prototype Vuilleumier heat pump burner. (*Courtesy of Prof. S. Schulz and the University of Dortmund.*)

- The heat exchangers and regenerators did not contribute to the dead-space volume, except as accounted for in the assumed parametric values.
- The dead-space volume was distributed between the engine and refrigerator segments in proportion to their size, rather than in proportion to the volume of heat exchangers and regenerators needed to meet the required heat duty of each segment.
- There was no pressure loss resulting from fluid flow friction in the heat exchangers and regenerators.

Figure 8.16 Packaged prototype Vuilleumier heat pump. (*Courtesy of Tokyo Gas Company.*)

Because of these assumptions, the parametric analyses gave no guidance on how much dead space would be associated with heat exchangers or regenerators of the size necessary to handle the heat transfer loads involved. Design optimization analysis, described in this section, removes these assumptions and allows more realistic evaluation and comparison of the potential performance of integrated heat pump embodiments. It can also be used in optimizing preliminary hardware designs. Design optimization analysis shows how sensitive heat pump performance is to the sizes of the heat exchangers and regenerators required for practical operation. This type of analysis can be used to show that two loss-generating phenomena—dead space and flow resistance—substantially influence heat pump performance.

Effects of realistic heat-exchanger design

In addition to the constraints of imperfect heat transfer, real heat exchangers and regenerators add dead space to the heat pump and reduce its effectiveness. These unavoidable additional volumes influence each embodiment to a different degree. Generally, embodiments with higher pressure swings are more sensitive to the amount of dead space.

Analysis of Real Integrated Heat Pumps   141

Real heat exchangers and regenerators also generate flow friction losses. These losses are closely related to the size of the heat exchangers and regenerators. Certain integrated heat pumps are more likely to have higher flow friction losses than others. The flow friction losses become a substantial part of the heat pump energy requirements when they become the source of additional auxiliary power requirements.

The parametric analyses described in Chaps. 6 and 7 did not take into account realistic heat-exchanger and regenerator designs. Additional calculation complexity is required to evaluate the required heat transfer area, regenerator size, corresponding dead-space volumes, and flow friction losses. The parametric evaluation allowed us to judge the performance of the individual concepts and their size sensitivity to these losses, but it did not allow a realistic evaluation of the maximum specific cooling and heating capacity and the effectiveness of the various embodiments. For example, the power required to overcome internal fluid friction cannot be evaluated by parametric analysis. The sensitivity of performance to various design parameters does not tell how practical the concept could be or how to make the necessary design compromises. For that reason, we supplemented the previous evaluation by specific calculations which show examples of a simple, but effective way to optimize the design.

Design optimization tools

Detailed computer programs for design of specific integrated heat pump embodiments have been developed by various agencies. For example, programs available from Professor Eder effectively evaluate all possible losses in the heat pumps he is developing. In contrast, the type of analysis presented in this chapter is less detailed, but it allows comparison of different embodiments. Calculated results presented later in the chapter illustrate the importance of design optimization analysis by comparing thermal-compression Vuilleumier and Stirling-Stirling (piston-piston) embodiments. These models are exemplified by a computer program for thermal-compression Vuilleumier embodiments, presented in Appendix F.

To compare the embodiments consistently, a common set of design specifications was selected. The heat pump operating parameters are given in Table 8.3. Table 8.4 gives the design specifications for the heat exchangers. This analysis was based on finned-tube cross-flow heat exchangers with characteristics as described by Kays and London.[15] Reasonable values for nominal outlet-temperature differences and helium-side Reynolds numbers are specified. The program calculates the required frontal area of the heat exchangers and the flow friction losses. Although higher Reynolds numbers would lead to

TABLE 8.3 Common Performance Parameters for the Integrated Heat Pumps Analyzed in Chap. 8

Working fluid	Helium
Cooling capacity	35 kW
Frequency	16.67 Hz
Mean pressure	5.0 MPa
Hot-side heat input temperature	547.8°C (1018°F)
Hot-side heat rejection temperature	65.6°C (150°F)
Cold-side heat input temperature	0°C (32°F)
Cold-side heat rejection temperature	65.6°C (150°F)

1 kW = 3413 Btu/h, 1 MPa = 145 psia.

TABLE 8.4 Heat-Exchanger Design Parameters for the Integrated Heat Pumps Analyzed in Chap. 8

	All heat exchangers
Material	Aluminum
Configuration	Finned-tube, cross-flow
	Hot heat exchanger
Temperature difference	30.0°C (54°F)
Reynolds number	5000
	Warm/hot heat exchanger
Temperature difference	6.0°C (10.8°F)
Reynolds number	8000
	Cold heat exchanger
Temperature difference	6.0°C (10.8°F)
Reynolds number	8000
	Warm/cold heat exchanger
Temperature difference	6.0°C (10.8°F)
Reynolds number	8000

better heat transfer, this would be at the expense of greater losses from fluid flow friction. Table 8.5 specifies the design parameters for the regenerators. Reasonable values for materials properties, geometry, and efficiency are specified. The required regenerator size and flow friction losses are calculated, again based on Kays and London's data.

TABLE 8.5 Regenerator Design Parameters for the Integrated Heat Pumps Analyzed in Chap. 8

Material	Stainless steel
Porosity	80%
Density	1538 kg/m^3 (95.9 lb/ft^3)
Length-to-diameter ratio	5.0
Efficiency	95%

Although the characteristics of the heat exchangers and regenerators are realistic, this design optimization analysis did not include alternative designs that would have better heat transfer and pressure loss characteristics. Also, there was no attempt to balance the competing effects of heat transfer effectiveness and pressure loss to optimize heat pump performance. However, programs of the type shown in Appendix F can be used for this balancing, which is necessary for effective prototype design.

While the parameters given in Tables 8.3, 8.4, and 8.5 were used in the evaluation of both of the embodiments analyzed, the working volume ratios defined below were optimized separately for each embodiment. This assured the maximum specific cooling capacity for each embodiment for the specified temperature limits and mean operating pressure. The optimization process is necessary because the working volume ratio, or ratio of the swept volume in the compression (hot) space to the swept volume in the expansion (cold) space, significantly affects the specific cooling capacity of an integrated heat pump. The computer program in Appendix F does not perform this optimization automatically. The user must seek the optimum by adjusting the working volume ratio and executing the program to produce results that are successively closer to optimal. For a given ratio of temperatures in the compression and expansion spaces, phase angle, and ratio of the dead-space volume to the total working volume, there is an optimum working volume ratio. This optimum ratio is calculated through iterative manual use of the program. The first step of this iteration is carried out for a set of specific temperatures (this parameter has the strongest influence on the specific cooling capacity), a phase angle of 90°, and an assumed dead space volume ratio of 0. The optimum working volume ratio for each cycle was found by running the program several times with different ratios to find the ratio which maximizes the specific cooling capacity of the heat pump. Through experience, we have learned that a phase angle of 90° produces results that are near enough to optimum that it is not necessary to optimize the design with respect to this variable.

The optimum working volume ratios found in this way are:

V_{hot}/V_{cold} (Vuilleumier)	1.3
$V_{compression}/V_{expansion}$ (Stirling refrigerator)	1.33
$V_{compression}/V_{expansion}$ (Stirling engine)	0.84

The computer program generates the following output data:

- COP_c
- Heat input needed to produce required cooling capacity

- Flow friction losses
- Dead-space volume
- Working volume
- Heat-exchanger characteristics
- Regenerator characteristics
- Specific cooling capacity

The program for design optimization of a thermal-compression Vuilleumier embodiment is given in Appendix F. The program starts by accepting input of the thermodynamic operating conditions, the required cooling capacity of an equivalent ideal-cycle heat pump, and the specified design parameters of the heat exchangers and regenerators. It also requires input of the expected hot-space volume and the ratio of the hot-space volume to the cold-space volume previously obtained from the parametric analysis. A good choice of expected hot-space volume will significantly reduce the number of iterations needed to obtain the final result.

The cold-space volume is calculated from the working volume ratio. At this point, the dead-space volume is assumed to be zero. The mass of working fluid, its flow, and the energy transfer required in regenerators and heat exchangers are calculated from the mean cycle pressure. Then, cooling capacity and cooling COP are calculated and compared with the COP from the previous iteration. During the first iteration, the program calculates the size of the heat exchangers and regenerators needed for heat transfer in and out, and within the cycle. Associated dead-space volume and flow friction losses are also calculated and added to the total volume of the machine and its energy balance. Then the calculation is repeated using the new dead-space volume. The addition of dead space changes the cooling capacity, COP_c, and regenerator and heat-exchanger loads. This calculation is repeated until the COP_c changes less than 1 percent from one iteration to the next. Then the calculated cooling capacity is compared with the input value for the desired cooling capacity. If the two capacities do not match, the program adjusts the value of the hot space volume and repeats the calculations. When calculated cooling capacities converge to within 1 percent, the program prints the results.

Tables 8.6 and 8.7 show the results of such calculations for the realistic heat exchanger and regenerator characteristics specified in Tables 8.3 through 8.5. Table 8.6 is for a thermal-compression Vuilleumier embodiment. The results are compared with corresponding results obtained from the simpler parametric model outlined in Chap. 6 and presented in Chap. 7. The parametric analysis was run with the dead-space vol-

TABLE 8.6 Thermal-Compression Vuilleumier-Cycle Embodiment Evaluation (Design Parameters Given in Tables 8.3 through 8.5)

	Ideal parametric model	Design optimization model
Dead-space volume fraction, %	71.35	71.35
Total working volume, cm^3	42,551	45,698
Hot-space volume, cm^3	nd	7400
Cold-space volume, cm^3	nd	5690
Maximum pressure, MPa	5.35	5.25
Pressure ratio	1.14	1.10
Hot heat-exchanger load, kW	14.42	62.72
Hot heat-exchanger void volume, cm^3	nd	1830
Hot-warm heat-exchanger load, kW	14.42	62.72
Hot-warm heat-exchanger void volume, cm^3	nd	3,832
Cold heat-exchanger load, kW	35.00	17.36
Cold heat-exchanger void volume, cm^3	nd	12,078
Cold-warm heat-exchanger load, kW	35.00	17.36
Cold-warm heat-exchanger void volume, cm^3	nd	11,227
Hot regenerator load, kW	nd	970.68
Hot regenerator void volume, cm^3	nd	981
Cold regenerator load, kW	nd	358.84
Cold regenerator void volume, cm^3	nd	2658
Ideal-cycle cooling capacity, kW	35.00	35.00
Real-cycle cooling capacity, kW	na	17.36
Ideal-cycle heat input, kW	14.42	14.42
Real-cycle heat input, kW	na	62.72
Flow friction losses, kW	0	0.68
Cooling COP	2.43	0.28

nd—Not determined.
na—Not applicable.
1 kW = 3413 Btu/h, 1 cm^3 = 0.0610 in.^{3}, 1 MPa = 145 psia.

ume fraction identified by the design optimization analysis when using realistic heat exchangers. Therefore, Table 8.6 compares the results of both types of analysis for that dead-space volume fraction. Comparing the ideal parametric results with the design optimization results shows that the realistic heat exchangers cause a 50 percent reduction in both capacity and COP_c compared with the ideal values. The effect of realistic regenerators is to reduce COP_c further, to approximately 10 percent of Carnot COP_c. The impact of realistic regenerators can be seen by noting that the thermal storage requirement for the engine segment is 55 times larger than the cooling capacity. The storage requirement for the refrigerator segment is 20 times larger than the cooling capacity. Realistic effects cause the relative sizes of the hot and cold heat exchangers to be reversed. Whereas the ideal parametric results indicate that the cold heat exchanger needs to exchange more heat than the hot heat exchanger, the additional heat needed to overcome the inefficiencies results in the hot heat ex-

TABLE 8.7 Stirling-Stirling (Piston-Piston) Cycle Embodiment Evaluation (Design Parameters Given in Tables 8.3 through 8.5)

	Ideal parametric model		Design optimization model	
	Engine	Refrigerator	Engine	Refrigerator
Dead-space volume fraction, %	57	82	57	82
Total working volume, cm³	1643	36,819	3160	50,541
Expansion space volume, cm³	nd	nd	733	3945
Compression space volume, cm³	nd	nd	618	5218
Maximum pressure, MPa	7.03	5.69	6.29	5.36
Pressure ratio	1.98	1.29	1.58	1.15
Compression-space heat-exchanger load, kW	nd	nd	13.99	33.89
Compression-space heat-exchanger void volume, cm³	nd	nd	852	21,923
Expansion-space heat-exchanger load, kW	14.42	35.00	25.44	25.47
Expansion-space heat-exchanger void volume, cm³	nd	nd	741	17,725
Regenerator load, kW	nd	nd	115.63	195.96
Regenerator void volume, cm³	nd	nd	215	1729
Ideal-cycle cooling capacity, kW	na	35.00	na	35.00
Real-cycle cooling capacity, kW	na	na	na	25.47
Ideal-cycle heat input, kW	14.42	na	14.42	na
Real-cycle heat input, kW	na	na	25.44	na
Flow friction losses, kW	0	0	0.13	0.47
	Heat pump		Heat pump	
Cooling COP	2.43		1.00	

nd—Not determined.
na—Not applicable.
1 kW = 3413 Btu/h, 1 cm³ = 0.0610 in.³, 1 MPa = 145 psia.

changer transferring almost four times as much as the cold heat exchanger.

Table 8.7 is a similar comparison for a Stirling-Stirling embodiment. It is based on a computer program similar to the one for the thermal-compression Vuilleumier embodiment shown in Appendix F. The Stirling-Stirling design optimization analysis indicates that using realistic heat exchangers will lead to a dead-space volume fraction of over 50 percent. Comparing the ideal results with the design optimization model results shows a pattern similar to that for the thermal-compression Vuilleumier. The realistic cooling capacity is reduced to 73 percent of the ideal capacity by the use of real heat exchangers. The effect of realistic regenerators is to further reduce COP_c, to 41 percent of Carnot COP_c. The impact of realistic regenerators is less than that in a thermal-compression Vuilleumier because the Stirling does not need as much working fluid to achieve the same capacity. The thermal storage requirement for the regenerators is only 4 to 8 times the cooling capacity, instead of 20 or more.

In both cases there are substantial size and performance differences between the design optimization analysis and the parametric analysis. Unlike the parametric analyses, design optimization evaluates the effects of heat-exchanger volume and thermal effectiveness, unequal distribution of dead space between the engine and refrigerator segments, and fluid flow friction loss. All of these are significant determinants of heat pump performance. The added capabilities of design optimization analysis can provide a more thorough comparative evaluation of alternative integrated heat pump embodiments. The output data can also be used in the design of equipment based on these embodiments to help determine the expected size and weight of a heat pump concept.

The computer programs could be enhanced by adding subroutines to evaluate mechanical friction losses and heat losses to the surroundings.

Model validation

Table 8.8 compares the behavior of a thermal-compression Vuilleumier embodiment with the predictions of the design optimization model. The laboratory data are for a prototype heat pump built and tested at the Technical University of Munich.[7] The design and operating conditions of the prototype (pressure, frequency, and temperature) were used as inputs to the model, which then calculated optimum volumes, heating capacity, and COP_h.

In applying the model, the total swept volume was first set equal to that of the laboratory prototype. Through iterative calculations, the size of the heat exchangers was adjusted until the dead-space volume ratio matched that of the prototype. The results show that this computer model provides good prediction of real heating capacity and COP_h.

Conclusions

Design optimization analysis is a simple, yet accurate tool for preliminary design of integrated heat pumps. It shows the extreme importance of heat-exchanger and regenerator design. It is clear that ordinary heat exchangers add either so much dead space or fluid friction losses to these heat pumps that their performance becomes unattractive. Improvement of heat exchangers and regenerators must command central attention in the development of any integrated heat pump.

TABLE 8.8 Laboratory Test Data for One Operating Point of a Vuilleumier Heat Pump Designed for Residential Heating

	Design and experimental data	Comparative analytical data*
Design data		
Working fluid	Helium	Helium
Pressure, MPa	3.0 (fill-up)	3.06 (average working)
Frequency, rpm	750	750
Swept volume, cm^3	1000	1000
Total working volume, cm^3	2100	1995
Hot-to-cold volume ratio	1	1
Total void volume, cm^3	1100	995
Relative void volume, %	52	50
Hot-side temperature, °C	400	400
Hot/warm-side temperature, °C	48	48
Operating conditions		
Cold-side temperature, °C	15	15
Cold/warm-side temperature, °C	48	48
Design performance requirements		
Hot regenerator efficiency, %	Unknown	95
Cold regenerator efficiency, %	Unknown	95
Flow friction losses, W	Unknown	50
Performance results		
Heating capacity, kW	4.00	3.68
Heating COP	2.1	1.97

*Data iteratively matched.
1 kW = 3413 Btu/h, 1 cm^3 = 0.0610 in.^{3}, 1 MPa = 145 psia.

References

1. Daniels, A., and F. K. du Pré, "Miniature Refrigerators for Electronic Devices," *Philips Tech Revue*, 32:49–56, 1971.
2. Stolfi, F., and A. K. de Jonge, "Stirling Cryogenerators with Linear Drive," *Philips Tech Revue*, 42:1–10, 1985.
3. de Jonge, A. K., "A Small Free-Piston Stirling Refrigerator," paper 799245 presented at the 14th Intersociety Energy Conversion Engineering Conference, Boston, August 1979.
4. Eder, F. X., "Thermally Actuated Heat Pump," paper presented at International Institute of Refrigeration Joint Meeting of Commissions B1, B2, E1, E2, University of Essen, West Germany, Sept. 7–9, 1981.
5. Private communication with Professor F. X. Eder, German text, July 1989.
6. Thomas, B., H. D. Kühl, and S. Schulz, "A Short-Cut Optimization of the Swept Volume Ratio for Regenerative Cycles," paper 899377 presented at the 24th Intersociety Energy Conversion Engineering Conference, Washington D. C., Aug. 6–11, 1989.
7. Eder, F. X., J. Blumenberg, W. Becker, M. Neubronner, A. Stübner, W. Messerschmidt, and A. Müller, "Der Vuilleumier-prozess als Wärmepumpe und Kältemaschine; analytische Behandlung und Messergebnisse (the Vuilleumier process as heat pump and cooling machine, analytical treatment and experimental re-

sults)," presented at the DKV 1989 Annual Meeting, Hanover, FRG, Nov. 22–24, 1989.
8. Beale, W. T., "Free-Piston Stirling Engines—Some Model Tests and Simulations," SAE Paper 690230, 1969.
9. Berchowitz, D. M., "The Design, Development and Performance of a Duplex Stirling Natural Gas Liquefier," *Proceedings of the 17th Intersociety Energy Conversion Engineering Conference,* Los Angeles, August 1982.
10. Gedeon, D., B. Penswick, and W. Beale, "Duplex Stirling Heating-Only Gas Fired Heat Pump—Phase II," final report, Gas Research Institute, March 1983.
11. Penswick, L. B., and I. Urieli, "Duplex Stirling Machines," paper 849045 presented at the 19th Intersociety Energy Conversion Engineering Conference, San Francisco, August 1984.
12. Kühl, H. D., N. Richter, and S. Schulz, "Computer Simulation of a Vuilleumier Cycle Heat Pump for Domestic Use," Paper 869125 presented at the 21st Intersociety Energy Conversion Engineering Conference, San Diego, August 25–29, 1986.
13. Richter, N., "Theoretische Untersuchungen und konstruktive Vorschläge für die Realisierung einer Vuilleumier-Wärmepumpe," doctoral thesis, Düsseldorf: VDI Verlag, 1988.
14. "Sanyo Develops World's First Non-Flon Gas Combustion Direct Drive Heat Pump 'DDHP' System," *JARN* (Japanese Refrigeration News), July 25, 1989, p. 9.
15. Kays, W. M. and A. L. London, *Compact Heat Exchangers,* New York: McGraw-Hill, 1984.

Appendix

Thermal-Compression Vuilleumier Heat Pump Program

The VUILL.BAS program, written in QuickBASIC version 4.0 (a trademark of Microsoft Corporation), simulates the ideal, isothermal operation of the traditional thermal-compression Vuilleumier heat pump.

The program works by determining the volume variations for each of the four spaces for one degree increments through one complete cycle of operation. It then determines the pressure variations through one cycle. Using the ratio of the computed average value to the design value for the mean operating pressure, the program adjusts the pressures for each cycle step to meet the design mean operating pressure. It also adjusts the amount of working fluid in the heat pump to correspond to the thermodynamic condition in each space.

Before executing, values of the following parameters must be specified within the program:

Parameter name	Parameter definition	Units
nt	Starting value for total quantity of working fluid	gram-moles
ttlvol	Total volume of all working spaces	cm^3
vh.vc	Ratio between the hot-space volume (space 1) and the cold-space volume (space 4); the warm-space volumes (spaces 2 and 3) are also calculated from this and the total volume	None
frc.ds	Fraction of the total volume designated as dead-space volume	None
aop	Average (mean) operating pressure	MPa
th	High temperature (hot space)	°F
ti	Intermediate temperature (warm spaces)	°F
tc	Low temperature (cold space)	°F

The above parameters are contained in the section marked *parameter initialization*. The program calculates the heat exchanged with the environment, the mass transfer between spaces, and the heat storage capacity of the regenerator assuming 100% effective operation. Unlike the other programs in this book, VUILL.BAS does not calculate the work exchanged with the environment, because this cycle neither produces nor uses external work. If the work were computed, it would be zero.

In thermal-compression Vuilleumier heat pumps, the working fluid automatically redistributes itself among the four spaces based on their volumes and temperatures. As a result, this program will calculate the performance for a heat pump with almost any realistic value for vh.vc, the hot-space volume to cold-space volume ratio. However, using different values of vh.vc will produce different capacities for the same total volume. There is an optimum ratio which produces the largest specific capacity. By running the simulation with different values of vh.vc, the user can determine the highest specific capacity.

Once the optimum value of vh.vc has been determined, the total working volume required to meet the design capacity is calculated by multiplying the specified initial working volume by the ratio of the design value of the cooling capacity to the calculated optimum capacity.

During execution, the program displays summary values for the simulation. It also displays a counter indicating calculation progress. The summary results consist of:

- Volume of each space at the 0, 90, 180, and 270° cycle steps, cm^3
- Pressure at the 0, 90, 180, and 270° cycle steps, MPa
- Maximum, minimum, and average pressures, MPa
- Heat into each space, J
- Heat out of each space, J
- Net heat transfer between each space and the environment, J
- Quantity of working fluid that is transferred between the hot and warm spaces (1 and 2), gram-moles
- Quantity of working fluid that is transferred between the warm and cold spaces (3 and 4), gram-moles
- Regenerator storage capacity needed to achieve 100% effectiveness (equivalent to the assumption of no heat transfer between spaces due to mass transfer), J

In addition to displaying these results on the monitor, the program stores these values in an ASCII text file named VUILL.DAT. Values for volumes, pressures, heat transfer between each space and the environment, and mass transfer between spaces are stored in 1° steps for the last complete operating cycle in VUILL.DAT.

154 Appendix A

```
'**************************************************************
' VUILL.BAS - simulation of a thermal compression Vuilleumier
' traditional configuration of two displacers in two cylinders
' hot and warm spaces in one cylinder
' cold and warm spaces in the other cylinder
'**************************************************************

'**************************************************************
' set flag for dynamic arrays
' $DYNAMIC

'**************************************************************
' define variable types
DEFDBL A-H
DEFDBL K-Z
DEFINT I-J

'**************************************************************
' array definition
'**************************************************************
' volume for each working space (hot, warm/hot, warm/cold, and cold),
' at cycle steps of one degree
DIM vol(4, 360)
' pressure, at one degree steps
DIM pres(360)
' working fluid in each working space, at one degree steps
DIM n(4, 360)
' absolute temperature in each of four working spaces
DIM t(4)
' heat into each space, over each degree step
DIM hi(4, 360)
' heat out of each space, over each degree step
DIM ho(4, 360)
' total heat into each space
DIM hit(4)
' total heat out of each space
DIM hot(4)
' net heat transfer of each space
DIM hn(4)

'**************************************************************
' constants
'**************************************************************
pi = 3.1415926#
bl$ = SPACE$(40)
' ideal gas constant, cm^3 MPa / K mol
r = 8.3143

'**************************************************************
' parameter initialization
'**************************************************************
' initial value for total amount of working fluid, moles
nt = 1
' total volume, cm^3
```

```
ttlvol = 1000
' calculated volume ratio between hot/warm cylinder volume and
' warm/cold cylinder volume
vh.vc = 2
' dead space volume fraction
frc.ds = .001
' desired average (mean) operating pressure, MPa
aop = 5
' hot temperature, intermediate temperature, cold temperature, degrees F
th = 1000
ti = 150
tc = 32

'*************************************************************
' computed working values
'*************************************************************
' convert temperatures to kelvin
t(1) = (th + 460) / 1.8
t(2) = (ti + 460) / 1.8
t(3) = (ti + 460) / 1.8
t(4) = (tc + 460) / 1.8
' compute active space volume fraction (no dead space is included)
frc.as = 1 - frc.ds
' one-half maximum volume for space 1 (hot)
vh.2 = ttlvol * vh.vc / (vh.vc + 1) / 2
' one-half maximum volume for space 4 (cold)
vc.2 = (ttlvol - (vh.2 * 2)) / 2

'*************************************************************
' output operation
'*************************************************************
' record initial values of analysis
CLS
OPEN "vuill.dat" FOR OUTPUT AS 1
PRINT "Thermal Compression Vuilleumier -- " + DATE$ + " | " + TIME$
PRINT #1, "Thermal Compression Vuilleumier -- " + DATE$ + " | " + TIME$

fmt$ = "Total #####.# cm^3.   vh/vc=##.###"
PRINT USING fmt$; ttlvol; vh.vc
PRINT #1, USING fmt$; ttlvol; vh.vc

fmt$ = "th=#### F    ti=#### F    tc=#### F    dead spc=#.###"
PRINT USING fmt$; th; ti; tc; frc.ds
PRINT #1, USING fmt$; th; ti; tc; frc.ds

'*************************************************************
' calculate volumes for each space for each degree step
'*************************************************************
' step through one complete cycle, with i = 0 to 360 degrees
+--FOR i = 0 TO 360
|   ' trace information
|       LOCATE , 1
|       PRINT "working on volume"; i;
|
```

Appendix A

```
|   ' calculate instantaneous volume for each space,
|   ' accounting for only active space
|      vol(1, i) = SIN(i * pi / 180) * vh.2 * frc.as + vh.2
|      vol(2, i) = vh.2 * 2 - vol(1, i)
|      vol(3, i) = SIN((i + 90) * pi / 180) * vc.2 * frc.as + vc.2
|      vol(4, i) = vc.2 * 2 - vol(3, i)
+--NEXT
   LOCATE , 1

   ' report on volume results
   fmt$ = "vol 1=###### cc   2=###### cc   3=###### cc   4=###### cc"
   PRINT USING fmt$; vol(1, 0); vol(2, 0); vol(3, 0); vol(4, 0)
   PRINT USING fmt$; vol(1, 90); vol(2, 90); vol(3, 90); vol(4, 90)
   PRINT USING fmt$; vol(1, 180); vol(2, 180); vol(3, 180); vol(4, 180)
   PRINT USING fmt$; vol(1, 270); vol(2, 270); vol(3, 270); vol(4, 270)
   PRINT #1, USING fmt$; vol(1, 0); vol(2, 0); vol(3, 0); vol(4, 0)
   PRINT #1, USING fmt$; vol(1, 90); vol(2, 90); vol(3, 90); vol(4, 90)
   PRINT #1, USING fmt$; vol(1, 180); vol(2, 180); vol(3, 180); vol(4, 180)
   PRINT #1, USING fmt$; vol(1, 270); vol(2, 270); vol(3, 270); vol(4, 270)
   PRINT #1, USING fmt$; vol(1, 360); vol(2, 360); vol(3, 360); vol(4, 360)

   '***********************************************************************
   ' calculate pressures at each degree step assuming uniform instantaneous
   ' pressure throughout
   '***********************************************************************
   ' step through one complete cycle, with i = 0 to 360 degrees
+--FOR i = 0 TO 360
|   ' trace information
|      LOCATE , 1
|      PRINT "working on pressure"; i;
|
|      pres(i) = r * nt / (vol(1, i) / t(1) + vol(2, i) / t(2) + vol(3, i) / t(3)
|         + vol(4, i) / t(4))
+--NEXT

   '***********************************************************************
   ' determine max, min, avg pressures
   '***********************************************************************
   pmax = pres(1)
   pmin = pres(1)
   pavg = pres(1)
   ' step through cycle
+--FOR i = 2 TO 360
|      IF pres(i) > pmax THEN pmax = pres(i)
|      IF pres(i) < pmin THEN pmin = pres(i)
|      pavg = pavg + pres(i)
+--NEXT
   pavg = pavg / 360

   '***********************************************************************
   ' adjust for desired mean operating pressure
   '***********************************************************************
   ' determine ratio between desired average operating pressure and
   ' computed average operating pressure
```

```
    rto = aop / pavg

    ' adjust total working fluid amount to reflect desired pressure
    ' change from computed value
    nt = nt * rto

    ' calculate new max, min and average values
    pmax = pmax * rto
    pmin = pmin * rto
    pavg = aop

    ' step through one complete cycle, with i = 0 to 360 degrees,
    ' calculating new pressure for each step based on change in
    ' amount of working fluid in moles
+--FOR i = 0 TO 360
|      pres(i) = pres(i) * rto
|   ' calculate amount of working fluid in each space for each degree step
|   ' j=1 == hot space; j=2 == warm/hot space;
|   ' j=3 == warm/cold space; j=4 == cold space
|   +--FOR j = 1 TO 4
|   |      n(j, i) = pres(i) * vol(j, i) / (r * t(j))
|   +--NEXT
+--NEXT

    ' record pressure at 90 degree steps
    LOCATE , 1
    fmt$ = "pres 0=##.## MPa    90=##.## MPa    180=##.## MPa    270=##.## MPa"
    PRINT USING fmt$; pres(0); pres(90); pres(180); pres(270)
    PRINT #1, USING fmt$; pres(0); pres(90); pres(180); pres(270)

    '••••••••••••••••••••••••••••••••••••••••••••••••••••••••••••••••••••
    ' print max, min and avg pressures
    '••••••••••••••••••••••••••••••••••••••••••••••••••••••••••••••••••••
    fmt$ = "pres max=##.## MPa   min=##.## MPa   avg=##.## MPa"
    PRINT USING fmt$; pmax; pmin; pavg
    PRINT #1, USING fmt$; pmax; pmin; pavg

    '••••••••••••••••••••••••••••••••••••••••••••••••••••••••••••••••••••
    ' calculate heat in and out
    '••••••••••••••••••••••••••••••••••••••••••••••••••••••••••••••••••••
    ' step through one complete cycle, with i = 0 to 360 degrees
+--FOR i = 1 TO 360
|   ' trace information
|      LOCATE , 1
|      PRINT "heat in/out"; i;
|
|   ' calculate average pressure between two degree steps
|      apres = (pres(i) + pres(i - 1)) / 2
|
|   ' step through each space
|   +--FOR j = 1 TO 4
|   |
|   'calculate heat based on PdV (average pressure * volume change)
|   |      ht = apres * (vol(j, i) - vol(j, i - 1))
```

Appendix A

```
|  |
|  ' determine whether heat is flowing in or out
|  ' heat in is a positive value; heat out is a negative value
|  |  +--IF ht > 0 THEN
|  |  |      hi(j, i) = ht
|  |  |      ho(j, i) = 0
|  |  +--ELSE
|  |  |      hi(j, i) = 0
|  |  |      ho(j, i) = ht
|  |  +--END IF
|  |
|  +--NEXT
+--NEXT

'*********************************************************************
' calculate total heat into and out of each space for one complete cycle
'*********************************************************************
   hit(1) = 0
   hit(2) = 0
   hit(3) = 0
   hit(4) = 0
   hot(1) = 0
   hot(2) = 0
   hot(3) = 0
   hot(4) = 0
   hn(1) = 0
   hn(2) = 0
   hn(3) = 0
   hn(4) = 0

   ' step through one complete cycle, with i = 1 to 360 degrees
+--FOR i = 1 TO 360
|  ' step through each space: hot, warm/hot, warm/cold, cold
|  +--FOR j = 1 TO 4
|  |  ' total heat into each space
|  |      hit(j) = hit(j) + hi(j, i)
|  |  ' total heat out of each space
|  |      hot(j) = hot(j) + ho(j, i)
|  |  ' total net heat (into and out of each space)
|  |      hn(j) = hn(j) + hi(j, i) + ho(j, i)
|  +--NEXT
+--NEXT

'*********************************************************************
' calculated results of heat transferred
'*********************************************************************
   LOCATE , 1
   fmt$ = "heat in  1=######.## J  2=######.## J  3=######.## J  4=######.## J"
   PRINT USING fmt$; hit(1); hit(2); hit(3); hit(4)
   fmt$ = "heat out 1=######.## J  2=######.## J  3=######.## J  4=######.## J"
   PRINT USING fmt$; hot(1); hot(2); hot(3); hot(4)
   fmt$ = "net heat 1=######.## J  2=######.## J  3=######.## J  4=######.## J"
   PRINT USING fmt$; hn(1); hn(2); hn(3); hn(4)
```

Thermal-Compression Vuilleumier Heat Pump Program 159

```
'********************************************************************
' calculate mass flow between spaces
'********************************************************************
    n1i = 0
    n1o = 0
    n2i = 0
    n2o = 0
    n3i = 0
    n3o = 0
    n4i = 0
    n4o = 0

    ' step through one complete cycle, with i = 1 to 360 degrees
+--FOR i = 1 TO 360
|   ' determine how much working fluid moved between previous and current steps
|   ' for space 1 (hot)
|       n1 = n(1, i) - n(1, i - 1)
|   ' determine direction of flow
|   +--IF n1 >= 0 THEN
|   |       n1i = n1i + n1
|   +--ELSE
|   |       n1o = n1o - n1
|   +--END IF
|
|   ' determine how much working fluid moved between previous and current steps
|   ' for space 4 (cold)
|       n4 = n(4, i) - n(4, i - 1)
|   ' determine direction of flow
|   +--IF n4 >= 0 THEN
|   |       n4i = n4i + n4
|   +--ELSE
|   |       n4o = n4o - n4
|   +--END IF
+--NEXT

    ' mass flow average (in and out) for hot space (to reduce numerical errors)
    n1 = (n1i + n1o) / 2
    ' mass flow average (in and out) for cold space (to reduce numerical errors)
    n4 = (n4i + n4o) / 2
    ' calculate amount of heat exchanged as gas goes from hot to warm/hot
    hs1 = n1 * 4 * 5.19 * (t(1) - t(2))
    ' calculate amount of heat exchanged as gas goes from warm/cold to cold
    hs4 = n4 * 4 * 5.19 * (t(3) - t(4))

    ' print results
    PRINT USING "mass transfer in/out (1-2): #####.####"; n1;
    PRINT USING "      (3-4): #####.####"; n4
    PRINT #1, USING "mass transfer in/out (1-2): #####.####"; n1;
    PRINT #1, USING "      (3-4): #####.####"; n4

    PRINT USING "regenerator storage (1-2): ######.# J"; hs1;
    PRINT USING "      (3-4): ######.# J"; hs4
    PRINT #1, USING "regenerator storage (1-2): ######.# J"; hs1;
    PRINT #1, USING "      (3-4): ######.# J"; hs4
```

Appendix A

```
'••••••••••••••••••••••••••••••••••••••••••••••••••••••••••••••••••••
' print out details for file
'••••••••••••••••••••••••••••••••••••••••••••••••••••••••••••••••••••
  PRINT #1, "volumes throughout cycle"
+--FOR i = 1 TO 360
|     PRINT #1, USING "###    #####.## cm^3"; i; vol(1, i);
|     PRINT #1, USING "    #####.## cm^3"; vol(2, i);
|     PRINT #1, USING "    #####.## cm^3"; vol(3, i);
|     PRINT #1, USING "    #####.## cm^3"; vol(4, i)
+--NEXT

  PRINT #1, "pressure throughout cycle"
+--FOR i = 1 TO 360
|     PRINT #1, USING "###    ###.## MPa"; i; pres(i)
+--NEXT

  PRINT #1, "heat transfer in/out of spaces"
+--FOR i = 1 TO 360
|     PRINT #1, USING "###    ####.#### J"; i; hi(1, i) + ho(1, i);
|     PRINT #1, USING "    ####.#### J"; hi(2, i) + ho(2, i);
|     PRINT #1, USING "    ####.#### J"; hi(3, i) + ho(3, i);
|     PRINT #1, USING "    ####.#### J"; hi(4, i) + ho(4, i)
+--NEXT

  PRINT #1, "mass transfer in/out of spaces"
+--FOR i = 1 TO 360
|     PRINT #1, USING "###"; i;
|   +--FOR j = 1 TO 4
|   |    PRINT #1, USING "    ##.#####"; n(j, i) - n(j, i - 1);
|   +--NEXT
|     PRINT #1,
+--NEXT

  ' close file
  CLOSE
  END
```

Appendix

B

Stirling-Stirling (Piston-Piston) Heat Pump Programs

The STRL-REF.BAS and STRL-ENG.BAS programs, written in QuickBASIC version 4.0 (a trademark of Microsoft Corporation), simulate the operation of an ideal, isothermal Stirling refrigerator segment and Stirling engine segment, respectively, each using the alpha (piston-piston) configuration. Each program simulates half of a Stirling-Stirling heat pump concept. To simulate a Stirling-Stirling heat pump, STRL-REF.BAS is run first to determine the amount of net work needed by the refrigerator. STRL-ENG.BAS is then run to match the net amount of work produced by the engine to the work needed by the refrigerator.

Both programs work by determining the volume variations for their two respective spaces for one degree increments through one complete cycle of operation. They then determine the pressure variations through one cycle. Using the ratio of the computed average value to the design value for the mean operating pressure, the programs adjust the pressures for each cycle step to meet the design mean operating pressure. They also adjust the amount of working fluid in the Stirling engine and Stirling refrigerator to correspond to the amount needed to produce the design pressure.

Before executing either program, values of the following parameters must be specified within the STRL-REF.BAS and STRL-ENG.BAS programs:

Parameter name	Parameter definition	Units
nt	Starting value for total quantity of working fluid	gram-moles
ttlvol	Total volume of all working spaces	cm^3
frc.ds	Fraction of the total volume designated as dead space	None

Parameter name	Parameter definition	Units
aop	Average (mean) operating pressure	MPa
th	High temperature (hot space) (for STRL-ENG.BAS only)	°F
ti	Intermediate temperature (warm space)	°F
tc	Low temperature (cold space) (for STRL-REF.BAS only)	°F

The above parameters are contained in the section marked *parameter initialization* within each program. Each program calculates the heat exchanged with the environment, the mass transfer between spaces, the heat storage capacity of the regenerator assuming 100% effective operation, and the work flow into and out of each simulated segment and net work for that segment for one cycle.

To simulate a Stirling-Stirling heat pump, STRL-REF.BAS must first be run. Its output includes the expected delivered cooling capacity for the working volume assigned to ttlvol. To compute the volume needed to produce the design capacity, ttlvol is multiplied by the ratio of the design value of the cooling capacity to the computed capacity. To determine the net work required, the net work input reported by the program must also be multiplied by this ratio.

STRL-ENG.BAS is run next to determine the net work output of the engine segment. Because the engine segment delivers the work that the refrigerator segment requires to operate, the size of the Stirling engine must be adjusted by the ratio of the calculated work required by the refrigerator, previously computed by STRL-REF.BAS, to the calculated value of work delivered, as computed by STRL-ENG.BAS. If additional work is required for another function or is being provided by another means, then the work required by the refrigerator must be adjusted by this value before the size of the engine segment can be computed.

During execution, each program displays summary values for the simulation. While calculating, they also display a counter indicating their progress. The summary results consist of:

- Maximum and minimum total working volume, cm^3
- Volume of each space at the 0, 90, 180, and 270° cycle steps, cm^3
- Pressure at the 0, 90, 180, and 270° cycle steps, MPa
- Maximum, minimum, and average pressures, MPa
- Heat into each space, J
- Heat out of each space, J
- Net heat transfer between each space and the environment, J

- Regenerator storage capacity needed for 100% effectiveness (equivalent to the assumption of no heat transfer between spaces due to mass transfer), J
- Quantity of working fluid that is transferred between the hot and warm spaces (1 and 2; for STRL-ENG.BAS only), gram-moles
- Work produced by the engine segment that must be delivered to the refrigerator segment (for STRL-ENG.BAS only), J
- Quantity of working fluid that is transferred between the warm and cold spaces (1 and 2; for STRL-REF.BAS only), gram-moles
- Work required by the refrigerator segment that must be input to run (for STRL-REF.BAS only), J

In addition to displaying these results on the monitor, each program stores these values in ASCII text files. STRL-REF.BAS uses STRL-REF.DAT, and STRL-ENG.BAS uses STRL-ENG.DAT. Values for volumes, pressures, heat transfer between each space and the environment, mass transfer between spaces, and work transfer between the cycle and the environment are stored in 1° steps for the last complete operating cycle in STRL-REF.DAT and STRL-ENG.DAT.

Appendix B

```
'*************************************************************
' STRL-REF.BAS - simulation of a Stirling cycle refrigerator in
' an alpha (piston-piston) configuration
'*************************************************************

'*************************************************************
' set flag for dynamic arrays
' $DYNAMIC

'*************************************************************
' define variable types
DEFDBL A-H
DEFDBL K-Z
DEFINT I-J

'*************************************************************
' array definition
'*************************************************************
' volume for each working space (warm and cold)
' at cycle angle steps of one degree
DIM vol(2, 360)
' pressure, at one degree steps
DIM pres(360)
' working fluid in each working space, at one degree steps
DIM n(2, 360)
' absolute temperature in each of three working spaces
DIM t(2)
' heat into each space, over each degree step
DIM hi(2, 360)
' heat out of each space, over each degree step
DIM ho(2, 360)
' total heat into each space
DIM hit(2)
' total heat out of each space
DIM hot(2)
' net heat transfer of each space
DIM hn(2)

'*************************************************************
' constants
'*************************************************************
pi = 3.1415926#
bl$ = SPACE$(40)
' ideal gas constant, cm^3 MPa / K mol
r = 8.3143

'*************************************************************
' parameter initialization
'*************************************************************
' initial value for total amount of working fluid, moles
nt = 1
' total volume, cm^3
ttlvol = 3168.3
' dead space volume fraction
```

Stirling-Stirling (Piston-Piston) Heat Pump Programs   165

```
    frc.ds = .4
    ' desired average (mean) operating pressure, MPa
    aop = 5
    ' warm temperature, cold temperature, degrees F
    ti = 150
    tc = 32

'*********************************************************************
' computed working values
'*********************************************************************
    ' convert temperatures to kelvin
    t(1) = (tc + 460) / 1.8
    t(2) = (ti + 460) / 1.8
    ' compute active space volume fraction (no dead space is included)
    frc.as = 1 - frc.ds
    ' half of maximum volume for hot space
    vh.2 = ttlvol / 4

'*********************************************************************
' output operation
'*********************************************************************
    ' record initial values of analysis
    CLS
    OPEN "strl-ref.dat" FOR OUTPUT AS 1
    PRINT "Stirling refrigerator - piston/piston -- " + DATE$ + " | " + TIME$
    PRINT #1, "Stirling refrigerator - piston/piston -- " + DATE$ + " | " + TIME$

    fmt$ = "Vt=#####.# cm^3    ti=#### F    tc=#### F    deadspc=#.###"
    PRINT USING fmt$; ttlvol; ti; tc; frc.ds
    PRINT #1, USING fmt$; ttlvol; ti; tc; frc.ds

'*********************************************************************
' set up initial placeholders for maximum and minimum total volumes
'*********************************************************************
    vmax = -100000
    vmin = 100000

'*********************************************************************
' calculate volumes for each space for each degree step
'*********************************************************************
    ' step through one complete cycle, with i = 0 to 360 degrees
+--FOR i = 0 TO 360
|    ' trace information
|       LOCATE , 1
|       PRINT "working on volume"; i;
|
|    ' calculate instantaneous volume for each space,
|    ' accounting for only active space
|       vol(1, i) = vh.2 - vh.2 * SIN(i * pi / 180) * frc.as
|       vol(2, i) = vh.2 - vh.2 * SIN((i - 90) * pi / 180) * frc.as
|
|    ' calculate total instantaneous volume to find total max and min volumes
|       vsum = vol(1, i) + vol(2, i)
|       IF vmax < vsum THEN vmax = vsum
```

Appendix B

```
|    IF vmin > vsum THEN vmin = vsum
+--NEXT
   LOCATE , 1

   ' report on volume results
   PRINT USING "vmax=#####.##    vmin=#####.##"; vmax; vmin
   PRINT #1, USING "vmax=#####.##    vmin=#####.##"; vmax; vmin

   fmt$ = "vol 1=####### cc    2=####### cc"
   PRINT USING fmt$; vol(1, 0); vol(2, 0)
   PRINT USING fmt$; vol(1, 90); vol(2, 90)
   PRINT USING fmt$; vol(1, 180); vol(2, 180)
   PRINT USING fmt$; vol(1, 270); vol(2, 270)
   PRINT USING fmt$; vol(1, 360); vol(2, 360)
   PRINT #1, USING fmt$; vol(1, 0); vol(2, 0)
   PRINT #1, USING fmt$; vol(1, 90); vol(2, 90)
   PRINT #1, USING fmt$; vol(1, 180); vol(2, 180)
   PRINT #1, USING fmt$; vol(1, 270); vol(2, 270)
   PRINT #1, USING fmt$; vol(1, 360); vol(2, 360)

   '.....................................................................
   ' calculate pressures at each degree step assuming uniform instantaneous
   ' pressure throughout
   '.....................................................................
   ' step through one complete cycle, with i = 0 to 360 degrees
+--FOR i = 0 TO 360
|  ' trace information
|     LOCATE , 1
|     PRINT "working on pressure"; i;
|
|     pres(i) = r * nt / (vol(1, i) / t(1) + vol(2, i) / t(2))
+--NEXT

   '.....................................................................
   ' determine max, min, avg pressures
   '.....................................................................
   pmax = pres(1)
   pmin = pres(1)
   pavg = pres(1)
   ' step through cycle
+--FOR i = 2 TO 360
|     IF pres(i) > pmax THEN pmax = pres(i)
|     IF pres(i) < pmin THEN pmin = pres(i)
|     pavg = pavg + pres(i)
+--NEXT
   pavg = pavg / 360

   '.....................................................................
   ' adjust for desired mean operating pressure
   '.....................................................................
   ' determine ratio between desired average operating pressure and
   ' computed average operating pressure
   rto = aop / pavg
```

```
' adjust total working fluid amount to reflect desired pressure
' change from computed value
nt = nt * rto
' calculate new max, min and average values
pmax = pmax * rto
pmin = pmin * rto
pavg = aop

' step through one complete cycle, with i = 0 to 360 degrees,
' calculating new pressure based on ratio between calculated and
' design average
+--FOR i = 0 TO 360
|    pres(i) = pres(i) * rto
|  ' calculate amount of working fluid in each space for each degree step
|  ' j = 1 == warm space; j = 2 == cold space
|  +--FOR j = 1 TO 2
|  |    n(j, i) = pres(i) * vol(j, i) / (r * t(j))
|  +--NEXT
+--NEXT

' record pressure at 90 degree steps
  LOCATE , 1
  fmt$ = "pres  0=##.## MPa   90=##.## MPa   180=##.## MPa   270=##.## MPa"
  PRINT USING fmt$; pres(0); pres(90); pres(180); pres(270)
  PRINT #1, USING fmt$; pres(0); pres(90); pres(180); pres(270)

'*****************************************************************
' print max, min and avg pressures
'*****************************************************************
  fmt$ = "pres max=##.## MPa   min=##.## MPa   avg=##.## MPa"
  PRINT USING fmt$; pmax; pmin; pavg
  PRINT #1, USING fmt$; pmax; pmin; pavg

'*****************************************************************
' calculate heat in and heat out
'*****************************************************************
' step through one complete cycle, with i = 0 to 360 degrees
+--FOR i = 1 TO 360
|  ' trace information
|     LOCATE , 1
|     PRINT "heat in/out"; i;
|
|  ' calculate average pressure between two degree steps
|     apres = (pres(i) + pres(i - 1)) / 2
|
|  ' step through each space
|  +--FOR j = 1 TO 2
|  |
|  ' calculate heat based on PdV (average pressure * volume change)
|  |    ht = apres * (vol(j, i) - vol(j, i - 1))
|  |
|  ' determine whether heat is flowing in or out
|  ' heat in is a positive value; heat out is a negative value
|  |  +--IF ht > 0 THEN
```

Appendix B

```
|  |  |       hi(j, i) = ht
|  |  |       ho(j, i) = 0
|  |  +--ELSE
|  |  |       hi(j, i) = 0
|  |  |       ho(j, i) = ht
|  |  +--END IF
|  +--NEXT
+--NEXT

   '****************************************************************
   ' calculate total heat into and out of each space for one complete cycle
   '****************************************************************
     hit(1) = 0
     hit(2) = 0
     hot(1) = 0
     hot(2) = 0
     hn(1) = 0
     hn(2) = 0

     ' step through one complete cycle, with i = 1 to 360 degrees
+--FOR i = 1 TO 360
|    ' step through each space: warm, cold
|    +--FOR j = 1 TO 2
|    |   ' total heat into each space
|    |       hit(j) = hit(j) + hi(j, i)
|    |   ' total heat out of each space
|    |       hot(j) = hot(j) + ho(j, i)
|    |   ' total net heat (into and out of) each space
|    |       hn(j) = hn(j) + hi(j, i) + ho(j, i)
|    +--NEXT
+--NEXT

     ' report and store results in file
     LOCATE , 1

   '****************************************************************
   ' calculated results of heat transferred
   '****************************************************************
     fmt$ = "heat 1=######.## J  2=######.## J"
+--FOR i = 90 TO 360 STEP 90
|    PRINT USING fmt$; hi(1, i) + ho(1, i); hi(2, i) + ho(2, i)
|    PRINT #1, USING fmt$; hi(1, i) + ho(1, i); hi(2, i) + ho(2, i)
+--NEXT

     fmt$ = "temp      1=###### F      2=###### F   "
     PRINT USING fmt$; t(1) * 1.8 - 460; t(2) * 1.8 - 460
     PRINT #1, USING fmt$; t(1) * 1.8 - 460; t(2) * 1.8 - 460
     fmt$ = "heat in  1=######.## J  2=######.## J"
     PRINT USING fmt$; hit(1); hit(2)
     PRINT #1, USING fmt$; hit(1); hit(2)
     fmt$ = "heat out 1=######.## J  2=######.## J"
     PRINT USING fmt$; hot(1); hot(2)
     PRINT #1, USING fmt$; hot(1); hot(2)
     fmt$ = "net heat 1=######.## J  2=######.## J"
```

```
        PRINT USING fmt$; hn(1); hn(2)
        PRINT #1, USING fmt$; hn(1); hn(2)

        '•••••••••••••••••••••••••••••••••••••••••••••••••••••••••••••••••
        ' calculate mass flow between spaces
        '•••••••••••••••••••••••••••••••••••••••••••••••••••••••••••••••••
        n1i = 0
        n1o = 0

        ' step through one complete cycle, with i = 1 to 360 degrees
   +--FOR i = 1 TO 360
   |    ' determine how much working fluid moved between previous and current steps
   |    ' for space 1 (warm)
   |       n1 = n(1, i) - n(1, i - 1)
   |    ' determine direction of flow
   |   +--IF n1 >= 0 THEN
   |   |      n1i = n1i + n1
   |   +--ELSE
   |   |      n1o = n1o - n1
   |   +--END IF
   +--NEXT

        ' mass flow average (in and out) for warm space
        n1 = (n1i + n1o) / 2

        ' calculate amount of heat exchanged with regenerator
        ' as gas flows from warm to cold
        hs1 = n1 * 4 * 5.19 * (t(1) - t(2))

        ' print results
        PRINT USING "mass transfer in/out (1-2): #####.####"; n1
        PRINT #1, USING "mass transfer in/out (1-2): #####.####"; n1

        PRINT USING "regenerator storage (1-2): ######.# J"; hs1
        PRINT #1, USING "regenerator storage (1-2): ######.# J"; hs1

        '•••••••••••••••••••••••••••••••••••••••••••••••••••••••••••••••••
        ' calculate work in/out
        '•••••••••••••••••••••••••••••••••••••••••••••••••••••••••••••••••
        wi = 0
        wo = 0
        ' step through one complete cycle, with i = 1 to 360 degrees
   +--FOR i = 1 TO 360
   |    ' calculate work
   |       w = (hi(1, i) + hi(2, i) + ho(1, i) + ho(2, i))
   |    ' determine direction of flow
   |    ' work out of the concept is a positive value;
   |    ' work into the concept is a negative value
   |   +--IF w < 0 THEN
   |   |      wi = wi + w
   |   +--ELSE
   |   |      wo = wo + w
   |   +--END IF
   +--NEXT
```

```
' print results
PRINT USING "work out: ######.# J"; wo;
PRINT USING "     in: ######.# J"; wi;
PRINT USING " absorbed: ######.# J"; wo + wi
PRINT #1, USING "work out: ######.# J"; wo;
PRINT #1, USING "     in: ######.# J"; wi;
PRINT #1, USING " absorbed: ######.# J"; wo + wi


'•••••••••••••••••••••••••••••••••••••••••••••••••••••••••••••••••••••
' print out details for file
'•••••••••••••••••••••••••••••••••••••••••••••••••••••••••••••••••••••
   PRINT #1, "volumes throughout cycle"
+--FOR i = 1 TO 360
|      PRINT #1, USING "###    #####.## cm^3"; i; vol(1, i);
|      PRINT #1, USING "       #####.## cm^3"; vol(2, i)
+--NEXT

   PRINT #1, "pressure throughout cycle"
+--FOR i = 1 TO 360
|      PRINT #1, USING "###   ###.## MPa"; i; pres(i)
+--NEXT

   PRINT #1, "heat transfer in/out of spaces"
+--FOR i = 1 TO 360
|      PRINT #1, USING "###    ####.#### J"; i; hi(1, i) + ho(1, i);
|      PRINT #1, USING "       ####.#### J"; hi(2, i) + ho(2, i)
+--NEXT

   PRINT #1, "mass transfer in/out of spaces"
+--FOR i = 1 TO 360
|      PRINT #1, USING "###"; i;
|  +--FOR j = 1 TO 2
|  |      PRINT #1, USING "    ##.#####"; n(j, i) - n(j, i - 1);
|  +--NEXT
|      PRINT #1,
+--NEXT

   PRINT #1, "work transfer in/out during cycle"
+--FOR i = 1 TO 360
|      w = -(hi(1, i) + hi(2, i) + ho(1, i) + ho(2, i))
|      PRINT #1, USING "###    ######.#### J"; i; w
+--NEXT

   ' close file
   CLOSE
   END
```

Stirling-Stirling (Piston-Piston) Heat Pump Programs

```
'******************************************************************
' STRL-ENG.BAS - simulation of a Stirling cycle engine in an alpha
' (piston-piston) configuration
'******************************************************************

'******************************************************************
' set flag for dynamic arrays
' $DYNAMIC

'******************************************************************
' define variable types
DEFDBL A-H
DEFDBL K-Z
DEFINT I-J

'******************************************************************
' array definition
'******************************************************************
' volume for each working space (hot and warm)
' at cycle angle steps of one degree
DIM vol(2, 360)
' pressure, at one degree steps
DIM pres(360)
' working fluid in each working space, at one degree steps
DIM n(2, 360)
' absolute temperature in each of three working spaces
DIM t(2)
' heat into each space, over each degree step
DIM hi(2, 360)
' heat out of each space, over each degree step
DIM ho(2, 360)
' total heat into each space
DIM hit(2)
' total heat out of each space
DIM hot(2)
' net heat transfer of each space
DIM hn(2)

'******************************************************************
' constants
'******************************************************************
pi = 3.1415926#
bl$ = SPACE$(40)
' ideal gas constant, cm^3 MPa / K mol
r = 8.3143

'******************************************************************
' parameter initialization
'******************************************************************
' initial value for total amount of working fluid, moles
nt = 1
' total volume, cm^3
ttlvol = 819.4
' dead space volume fraction
```

172 Appendix B

```
    frc.ds = .4
    ' desired average (mean) operating pressure, MPa
    aop = 5
    ' hot temperature, warm temperature, degrees F
    th = 1000
    ti = 150

    '••••••••••••••••••••••••••••••••••••••••••••••••••••••••••••••••••
    ' computed working values
    '••••••••••••••••••••••••••••••••••••••••••••••••••••••••••••••••••
    ' convert temperatures to kelvin
    t(1) = (th + 460) / 1.8' kelvin
    t(2) = (ti + 460) / 1.8' kelvin
    ' compute active space volume fraction (no dead space is included)
    frc.as = 1 - frc.ds
    ' half of maximum volume for hot space
    vh.2 = ttlvol / 4

    '••••••••••••••••••••••••••••••••••••••••••••••••••••••••••••••••••
    ' output operation
    '••••••••••••••••••••••••••••••••••••••••••••••••••••••••••••••••••
    ' record initial values of analysis
    CLS
    OPEN "strl-eng.dat" FOR OUTPUT AS 1
    PRINT "Stirling engine - piston/piston -- " + DATE$ + " | " + TIME$
    PRINT #1, "Stirling engine - piston/piston -- " + DATE$ + " | " + TIME$

    fmt$ = "Vt=#####.#    th=#### F    ti=#### F    dead spc=#.###"
    PRINT USING fmt$; ttlvol; th; ti; frc.ds
    PRINT #1, USING fmt$; ttlvol; th; ti; frc.ds

    '••••••••••••••••••••••••••••••••••••••••••••••••••••••••••••••••••
    ' set up initial placeholders for maximum and minimum total volumes
    '••••••••••••••••••••••••••••••••••••••••••••••••••••••••••••••••••
    vmax = -100000
    vmin = 100000


    '••••••••••••••••••••••••••••••••••••••••••••••••••••••••••••••••••
    ' calculate volumes for each space for each degree step
    '••••••••••••••••••••••••••••••••••••••••••••••••••••••••••••••••••
    ' step through one complete cycle, with i = 0 to 360 degrees
+---FOR i = 0 TO 360
|   ' trace information
|       LOCATE , 1
|       PRINT "working on volume"; i;
|
|   ' calculate instantaneous volume for each space,
|   ' accounting for only active space
|       vol(1, i) = vh.2 - vh.2 * SIN(i * pi / 180) * frc.as
|       vol(2, i) = vh.2 - vh.2 * SIN((i - 90) * pi / 180) * frc.as
|
|   ' calculate total instantaneous volume to find total max and min volumes
|       vsum = vol(1, i) + vol(2, i)
|       IF vmax < vsum THEN vmax = vsum
```

Stirling-Stirling (Piston-Piston) Heat Pump Programs 173

```
 |      IF vmin > vsum THEN vmin = vsum
 +--NEXT
       LOCATE , 1

        ' report on volume results
        PRINT USING "vmax=#####.#     vmin=#####.#"; vmax; vmin
        PRINT #1, USING "vmax=#####.#     vmin=#####.#"; vmax; vmin

        fmt$ = "vol 1=####### cc    2=####### cc"
        PRINT USING fmt$; vol(1, 0); vol(2, 0)
        PRINT USING fmt$; vol(1, 90); vol(2, 90)
        PRINT USING fmt$; vol(1, 180); vol(2, 180)
        PRINT USING fmt$; vol(1, 270); vol(2, 270)
        PRINT #1, USING fmt$; vol(1, 0); vol(2, 0)
        PRINT #1, USING fmt$; vol(1, 90); vol(2, 90)
        PRINT #1, USING fmt$; vol(1, 180); vol(2, 180)
        PRINT #1, USING fmt$; vol(1, 270); vol(2, 270)

       '******************************************************
       ' calculate pressures at each degree step assuming uniform instantaneous
       ' pressure throughout
       '******************************************************
        ' step through one complete cycle, with i = 0 to 360 degrees
 +--FOR i = 0 TO 360
 |     ' trace information
 |        LOCATE , 1
 |        PRINT "working on pressure"; i;
 |
 |        pres(i) = r * nt / (vol(1, i) / t(1) + vol(2, i) / t(2))
 |
 +--NEXT

       '******************************************************
       ' determine max, min, avg pressures
       '******************************************************
        pmax = pres(1)
        pmin = pres(1)
        pavg = pres(1)

        ' step through cycle
 +--FOR i = 2 TO 360
 |      IF pres(i) > pmax THEN pmax = pres(i)
 |      IF pres(i) < pmin THEN pmin = pres(i)
 |      pavg = pavg + pres(i)
 +--NEXT
        pavg = pavg / 360

       '******************************************************
       ' adjust for desired mean operating pressure
       '******************************************************
        ' determine ratio between desired average operating pressure and
        ' computed average operating pressure
        rto = aop / pavg
```

Appendix B

```
    ' adjust total working fluid amount to reflect desired pressure
    ' change from computed value
    nt = nt * rto
    ' calculate new max, min and average values
    pmax = pmax * rto
    pmin = pmin * rto
    pavg = aop

    ' step through one complete cycle, with i = 0 to 360 degrees,
    ' calculating new instantaneous pressure based on ratio of design average
    ' to calculated average
+--FOR i = 0 TO 360
|     pres(i) = pres(i) * rto
|   ' calculate amount of working fluid in each space for each degree step
|   ' j = 1 == hot space; j = 2 == warm space
|   +--FOR j = 1 TO 2
|   |     n(j, i) = pres(i) * vol(j, i) / (r * t(j))
|   +--NEXT
+--NEXT

    ' record pressures at 90 degree steps
    LOCATE , 1
    fmt$ = "pres 0=##.## MPa    90=##.## MPa    180=##.## MPa    270=##.## MPa"
    PRINT USING fmt$; pres(0); pres(90); pres(180); pres(270)
    PRINT #1, USING fmt$; pres(0); pres(90); pres(180); pres(270)

    '••••••••••••••••••••••••••••••••••••••••••••••••••••••••••••••••••••
    ' print max, min and avg pressures
    '••••••••••••••••••••••••••••••••••••••••••••••••••••••••••••••••••••
    fmt$ = "pres max=##.## MPa    min=##.## MPa    avg=##.## MPa"
    PRINT USING fmt$; pmax; pmin; pavg
    PRINT #1, USING fmt$; pmax; pmin; pavg

    '••••••••••••••••••••••••••••••••••••••••••••••••••••••••••••••••••••
    ' calculate heat in and heat out
    '••••••••••••••••••••••••••••••••••••••••••••••••••••••••••••••••••••
    ' step through one complete cycle, with i = 0 to 360 degrees
+--FOR i = 1 TO 360
|   ' trace information
|       LOCATE , 1
|       PRINT "heat in/out"; i;
|
|   ' calculate average pressure between two degree steps
|       apres = (pres(i) + pres(i - 1)) / 2
|
|   ' step through each space
|   +--FOR j = 1 TO 2
|   |
|   |  ' calculate heat based on PdV (average pressure * volume change)
|   |      ht = apres * (vol(j, i) - vol(j, i - 1))
|   |
|   |  ' determine whether heat is flowing in or out
|   |  ' heat in is a positive value; heat out is a negative value
|   |  +--IF ht > 0 THEN
```

```
|  |  |        hi(j, i) = ht
|  |  |        ho(j, i) = 0
|  |  +--ELSE
|  |  |        hi(j, i) = 0
|  |  |        ho(j, i) = ht
|  |  +--END IF
|  |
|  +--NEXT
|
+--NEXT


  '*********************************************************
  ' calculate total heat into and out of each space for one complete cycle
  '*********************************************************
    hit(1) = 0
    hit(2) = 0
    hot(1) = 0
    hot(2) = 0
    hn(1) = 0
    hn(2) = 0

    ' step through one complete cycle, with i = 1 to 360 degrees
+--FOR i = 1 TO 360
|   ' step through each space, hot, warm
|   +--FOR j = 1 TO 2
|   |   ' total heat into each space
|   |        hit(j) = hit(j) + hi(j, i)
|   |   ' total heat out of each space
|   |        hot(j) = hot(j) + ho(j, i)
|   |   ' total net heat (into and out of each space)
|   |        hn(j) = hn(j) + hi(j, i) + ho(j, i)
|   +--NEXT
+--NEXT

    ' report and store results in file
    LOCATE , 1

  '*********************************************************
  ' calculated results of heat transferred
  '*********************************************************

    fmt$ = "temp      1=###### F      2=###### F    "
    PRINT USING fmt$; t(1) * 1.8 - 460; t(2) * 1.8 - 460
    PRINT #1, USING fmt$; t(1) * 1.8 - 460; t(2) * 1.8 - 460
    fmt$ = "heat in   1=######.## J  2=######.## J"
    PRINT USING fmt$; hit(1); hit(2)
    PRINT #1, USING fmt$; hit(1); hit(2)
    fmt$ = "heat out  1=######.## J  2=######.## J"
    PRINT USING fmt$; hot(1); hot(2)
    PRINT #1, USING fmt$; hot(1); hot(2)
    fmt$ = "net heat  1=######.## J  2=######.## J"
    PRINT USING fmt$; hn(1); hn(2)
    PRINT #1, USING fmt$; hn(1); hn(2)
```

Appendix B

```
'••••••••••••••••••••••••••••••••••••••••••••••••••••••••••••••••••••
' calculate mass flow between spaces
'••••••••••••••••••••••••••••••••••••••••••••••••••••••••••••••••••••
   n1i = 0
   n1o = 0

   ' step through one complete cycle, with i = 1 to 360 degrees
+--FOR i = 1 TO 360
|  ' determine how much working fluid moved between previous and current steps
|  ' for space 1 (hot)
|     n1 = n(1, i) - n(1, i - 1)
|  ' determine direction of flow
|  +--IF n1 >= 0 THEN
|  |     n1i = n1i + n1
|  +--ELSE
|  |     n1o = n1o - n1
|  +--END IF
+--NEXT

   ' mass flow average (in and out) for hot space (to reduce numerical errors)
   n1 = (n1i + n1o) / 2

   ' calculate amount of heat exchanged with regenerator
   ' as gas goes from hot to warm
   hs1 = n1 * 4 * 5.19 * (t(1) - t(2))

   ' print results
   PRINT USING "mass transfer in/out (1-2): #####.####"; n1
   PRINT #1, USING "mass transfer in/out (1-2): #####.####"; n1

   PRINT USING "regenerator storage (1-2): ######.# J"; hs1
   PRINT #1, USING "regenerator storage (1-2): ######.# J"; hs1

'••••••••••••••••••••••••••••••••••••••••••••••••••••••••••••••••••••
' calculate work in/out
'••••••••••••••••••••••••••••••••••••••••••••••••••••••••••••••••••••
   wi = 0
   wo = 0
   ' step through one complete cycle, with i = 1 to 360 degrees
+--FOR i = 1 TO 360
|  ' calculate work
|     w = (hi(1, i) + hi(2, i) + ho(1, i) + ho(2, i))
|  ' determine direction of flow
|  ' work out of the concept is a positive value;
|  ' work into the concept is a negative value
|  +--IF w < 0 THEN
|  |     wi = wi + w
|  +--ELSE
|  |     wo = wo + w
|  +--END IF
+--NEXT

   ' print results
   PRINT USING "work out: ######.# J"; wo;
```

```
    PRINT USING "    in: ######.# J"; wi;
    PRINT USING "   net: ######.# J"; wo + wi
    PRINT #1, USING "work out: ######.# J"; wo;
    PRINT #1, USING "    in: ######.# J"; wi;
    PRINT #1, USING "   net: ######.# J"; wo + wi

    '****************************************************************
    ' print out details for file
    '****************************************************************
    PRINT #1, "volumes throughout cycle"
+--FOR i = 1 TO 360
|       PRINT #1, USING "###   #####.## cm^3"; i; vol(1, i);
|       PRINT #1, USING "   #####.## cm^3"; vol(2, i)
+--NEXT

    PRINT #1, "pressure throughout cycle"
+--FOR i = 1 TO 360
|       PRINT #1, USING "###   ###.## MPa"; i; pres(i)
+--NEXT

    PRINT #1, "heat transfer in/out of spaces"
+--FOR i = 1 TO 360
|       PRINT #1, USING "###   ####.#### J"; i; hi(1, i) + ho(1, i);
|       PRINT #1, USING "   ####.#### J"; hi(2, i) + ho(2, i)
+--NEXT

    PRINT #1, "mass transfer in/out of spaces"
+--FOR i = 1 TO 360
|       PRINT #1, USING "###"; i;
|  +--FOR j = 1 TO 2
|  |       PRINT #1, USING "   ##.#####"; n(j, i) - n(j, i - 1);
|  +--NEXT
|       PRINT #1,
+--NEXT

    PRINT #1, "work transfer in/out during cycle"
+--FOR i = 1 TO 360
|       w = -(hi(1, i) + hi(2, i) + ho(1, i) + ho(2, i))
|       PRINT #1, USING "###   ######.#### J"; i; w
+--NEXT

    ' close file
    CLOSE
    END
```

Appendix C

Stirling-Stirling (Displacer-Piston) Heat Pump Programs

The STPD-REF.BAS and STPD-ENG.BAS programs, written in QuickBASIC version 4.0 (a trademark of Microsoft Corporation), simulate the operation of an ideal, isothermal Stirling refrigerator segment and Stirling engine segment, respectively, each using the beta (displacer-piston) configuration. To simulate a Stirling-Stirling heat pump, STPD-REF.BAS is run first to determine the amount of net work needed by the refrigerator. STPD-ENG.BAS is then run to match the net amount of work produced by the engine to the work needed by the refrigerator.

Both programs work by determining the volume variations for their two respective spaces for one degree increments through one complete cycle of operation. They then determine the pressure variations through one cycle. Using the ratio of the computed average value to the design value for the mean operating pressure, the programs adjust the pressures for each cycle step to meet the design mean operating pressure. They also adjust the amount of working fluid in the Stirling engine and Stirling refrigerator to correspond to the amount needed to produce the design pressure.

Before executing either program, values of the following parameters must be specified within the STPD-REF.BAS and STPD-ENG.BAS programs:

Parameter name	Parameter definition	Units
nt	Starting value for total quantity of working fluid	gram-moles
ttlvol	Total volume of all working spaces	cm^3

Parameter name	Parameter definition	Units
frc.ds	Fraction of the total volume designated as dead space	None
aop	Average (mean) operating pressure	MPa
th	High temperature (hot space) (for STPD-ENG.BAS only)	°F
ti	Intermediate temperature (warm space)	°F
tc	Low temperature (cold space) (for STPD-REF.BAS only)	°F

The above parameters are contained in the section marked *parameter initialization* within each program. Each program calculates the heat exchanged with the environment, the mass transfer between spaces, the heat storage capacity of the regenerator assuming 100% effective operation, and the work flow into and out of each simulated segment and net work for that segment for one cycle.

To simulate a Stirling-Stirling heat pump, STPD-REF.BAS must first be run. Its output includes the expected delivered cooling capacity for the working volume assigned to ttlvol. To compute the volume needed to produce the design capacity, ttlvol is multiplied by the ratio of the design value of the cooling capacity to the computed capacity. To determine the net refrigerator work required, the net work input reported by the program must also be multiplied by this ratio.

STPD-ENG.BAS is run next to determine the net work output of the engine segment. Since the engine segment delivers the work that the refrigerator segment requires to operate, the size of the Stirling engine must be adjusted by the ratio of the calculated work required by the refrigerator, previously computed by STPD-REF.BAS, to the calculated value of work delivered, as computed by STPD-ENG.BAS. If additional work is required for another function or is being provided by another means, then the work required by the refrigerator must be adjusted by this value before the size of the engine segment can be computed.

During execution, each program displays summary values for the simulation. While calculating, they also display a counter indicating their progress. The summary results consist of:

- Maximum and minimum total working volume, cm^3
- Volume of each space at the 0, 90, 180, and 270° cycle steps, cm^3
- Pressure at the 0, 90, 180, and 270° cycle steps, MPa

- Maximum, minimum, and average pressures, MPa
- Heat into each space, J
- Heat out of each space, J
- Net heat transfer between each space and the environment, J
- Regenerator storage capacity needed for 100% effectiveness (equivalent to the assumption of no heat transfer between spaces due to mass transfer), J
- Quantity of working fluid that is transferred between the hot and warm spaces (1 and 2; for STPD-ENG.BAS only), gram-moles
- Work produced by the engine segment that must be delivered to the environment (for STPD-ENG.BAS only), J
- Quantity of working fluid that is transferred between the warm and cold spaces (1 and 2; for STPD-REF.BAS only), gram-moles
- Work required by the refrigerator segment that must be input to run (for STPD-REF.BAS only), J

In addition to displaying these results on the monitor, each program stores these values in ASCII text files. STPD-REF.BAS uses STPD-REF.DAT, and STPD-ENG.BAS uses STPD-ENG.DAT. Values for volumes, pressures, heat transfer between each space and the environment, mass transfer between spaces, and work transfer between the cycle and the environment are stored in 1° steps for the last complete operating cycle in STPD-REF.DAT and STPD-ENG.DAT.

182   Appendix C

```
'*****************************************************************
' STPD-REF.BAS - simulation of a Stirling cycle refrigerator
' in a piston-displacer configuration
'*****************************************************************

'*****************************************************************
' set flag for dynamic arrays
' $DYNAMIC

'*****************************************************************
' define variable types
DEFDBL A-H
DEFDBL K-Z
DEFINT I-J

'*****************************************************************
' array definition
'*****************************************************************
' volume for each working space (warm and cold)
' at cycle angle steps of one degree
DIM vol(2, 360)
' pressure, at one degree steps
DIM pres(360)
' working fluid in each working space, at one degree steps
DIM n(2, 360)
' absolute temperature in each working space
DIM t(2)
' heat into each space, over each degree step
DIM hi(2, 360)
' heat out of each space, over each degree step
DIM ho(2, 360)
' total heat into each space
DIM hit(2)
' total heat out of each space
DIM hot(2)
' net heat transfer of each space
DIM hn(2)

'*****************************************************************
' constants
'*****************************************************************
pi = 3.1415926#
bl$ = SPACE$(40)
' ideal gas constant, cm^3 MPa / K mol
r = 8.3143

'*****************************************************************
' parameter initialization
'*****************************************************************
' initial value for total amount of working fluid, moles
nt = 1
' total volume, cm^3
ttlvol = 1000
' dead space volume fraction
```

```
    frc.ds = .001
    ' desired average (mean) operating pressure, MPa
    aop = 5
    ' warm temperature, cold temperature, degrees F
    ti = 150
    tc = 32


    '**********************************************************************
    ' computed working values
    '**********************************************************************
    ' convert temperatures to kelvin
    t(1) = (tc + 460) / 1.8  ' kelvin
    t(2) = (ti + 460) / 1.8  ' kelvin
    ' compute active space volume fraction (no dead space is included)
    frc.as = 1 - frc.ds

    '**********************************************************************
    ' output operation
    '**********************************************************************
    ' record initial values of analysis
    CLS
    OPEN "stpd-ref.dat" FOR OUTPUT AS 1
    PRINT "Stirling refrigerator - piston/displacer -- " + DATE$ + " | " + TIME$
    PRINT #1, "Stirling refrigerator - piston/displacer -- " + DATE$ + " | " +
      TIME$

    fmt$ = "ti=#### F   tc=#### F    dead spc=#.###"
    PRINT USING fmt$; ti; tc; frc.ds
    PRINT #1, USING fmt$; ti; tc; frc.ds


    '**********************************************************************
    ' calculate volumes for each space for each degree step
    '**********************************************************************
    vol1 = ttlvol * .5 * frc.as
    vol2 = vol1
    offset = 9999999

    ' step through one complete cycle, with i = 0 to 360 degrees
+--FOR i = 0 TO 360
|   ' trace information
|     LOCATE , 1
|     PRINT "working on volume"; i;
|     vol(1, i) = vol1 * (1 + SIN((i) * pi / 180)) / 2
|     vol(2, i) = vol2 * (1 + SIN((i - 90) * pi / 180)) / 2
|
|   ' search for maximum overlap between cold and warm spaces
|  +--IF (vol(2, i) - vol(1, i)) < offset THEN
|  |     offset = vol(2, i) - vol(1, i)
|  |     mi = i
|  +--END IF
+--NEXT
    LOCATE , 1

    ' determine an offsetting value to produce no overlap between
```

Appendix C

```
' warm and cold spaces
offset = -offset
delta = ttlvol * frc.as / (vol1 + offset)

' report on adjusted volume parameters
fmt$ = "ttl=#####.#  v1=#####.#  v2=#####.#  os=#####.#"
PRINT USING fmt$; ttlvol; vol1; vol2; offset
PRINT #1, USING fmt$; ttlvol; vol1; vol2; offset

v1l = 99999
v2l = 99999
vmin = 99999
v1h = 0
v2h = 0
vmax = 0

' adjust volumes allocated to each space by amount needed to shift
' piston away from displacer
+--FOR i = 0 TO 360
|     vol(1, i) = vol(1, i) * delta
|     vol(2, i) = (vol(2, i) + offset) * delta - vol(1, i) + vol2 * frc.ds
|     vol(1, i) = vol(1, i) + vol1 * frc.ds
|
|   ' calculate total instantaneous volume to find total max and min volumes
|     vsum = vol(1, i) + vol(2, i)
|     IF vmax < vsum THEN vmax = vsum
|     IF vmin > vsum THEN vmin = vsum
|     IF vol(1, i) > v1h THEN v1h = vol(1, i)
|     IF vol(1, i) < v1l THEN v1l = vol(1, i)
|     IF vol(2, i) > v2h THEN v2h = vol(2, i)
|     IF vol(2, i) < v2l THEN v2l = vol(2, i)
+--NEXT
   LOCATE , 1

   ' report on volume results
   PRINT USING "vmax=#####.#   vmin=#####.#"; vmax; vmin
   PRINT #1, USING "vmax=#####.#   vmin=#####.#"; vmax; vmin
   PRINT USING "min vol1=#####.#   vol2=#####.#"; v1l; v2l
   PRINT USING "max vol1=#####.#   vol2=#####.#"; v1h; v2h
   PRINT #1, USING "min vol1=#####.#   vol2=#####.#"; v1l; v2l
   PRINT #1, USING "max vol1=#####.#   vol2=#####.#"; v1h; v2h

   fmt$ = "vol 1=#######  cc    2=#######  cc   ### deg"
   PRINT USING fmt$; vol(1, 0); vol(2, 0); 0
   PRINT USING fmt$; vol(1, 45); vol(2, 45); 45
   PRINT USING fmt$; vol(1, 90); vol(2, 90); 90
   PRINT USING fmt$; vol(1, 135); vol(2, 135); 135
   PRINT USING fmt$; vol(1, 180); vol(2, 180); 180
   PRINT USING fmt$; vol(1, 225); vol(2, 225); 225
   PRINT USING fmt$; vol(1, 270); vol(2, 270); 270
   PRINT USING fmt$; vol(1, 315); vol(2, 315); 315
   PRINT #1, USING fmt$; vol(1, 0); vol(2, 0); 0
   PRINT #1, USING fmt$; vol(1, 45); vol(2, 45); 45
   PRINT #1, USING fmt$; vol(1, 90); vol(2, 90); 90
```

```
      PRINT #1, USING fmt$; vol(1, 135); vol(2, 135); 135
      PRINT #1, USING fmt$; vol(1, 180); vol(2, 180); 180
      PRINT #1, USING fmt$; vol(1, 225); vol(2, 225); 225
      PRINT #1, USING fmt$; vol(1, 270); vol(2, 270); 270
      PRINT #1, USING fmt$; vol(1, 315); vol(2, 315); 315

      '•••••••••••••••••••••••••••••••••••••••••••••••••••••••••••••••••
      ' calculate pressures at each degree step assuming uniform instantaneous
      ' pressure throughout
      '•••••••••••••••••••••••••••••••••••••••••••••••••••••••••••••••••
      ' step through one complete cycle, with i = 0 to 360 degrees
+--FOR i = 0 TO 360
|     ' trace information
|       LOCATE , 1
|       PRINT "working on pressure"; i;
|
|       pres(i) = r * nt / (vol(1, i) / t(1) + vol(2, i) / t(2))
|
|     ' end of loop
+--NEXT

      '•••••••••••••••••••••••••••••••••••••••••••••••••••••••••••••••••
      ' determine max, min, avg pressures
      '•••••••••••••••••••••••••••••••••••••••••••••••••••••••••••••••••
      pmax = pres(1)
      pmin = pres(1)
      pavg = pres(1)

      ' step through cycle
+--FOR i = 2 TO 360
|     IF pres(i) > pmax THEN pmax = pres(i)
|     IF pres(i) < pmin THEN pmin = pres(i)
|     pavg = pavg + pres(i)
+--NEXT
      pavg = pavg / 360

      '•••••••••••••••••••••••••••••••••••••••••••••••••••••••••••••••••
      ' adjust for desired mean operating pressure
      '•••••••••••••••••••••••••••••••••••••••••••••••••••••••••••••••••
      ' determine ratio between desired average operating pressure and
      ' computed average operating pressure
      rto = aop / pavg

      ' adjust total working fluid amount to reflect desired pressure
      ' change from computed value
      nt = nt * rto
      ' calculate new max, min and average values
      pmax = pmax * rto
      pmin = pmin * rto
      pavg = aop

      ' step through one complete cycle, with i = 0 to 360 degrees,
      ' calculating new pressure based on ratio between calculated and
      ' design average
```

186   Appendix C

```
+--FOR i = 0 TO 360
|    pres(i) = pres(i) * rto
|    ' calculate amount of working fluid in each space for each degree step
|    ' j = 1 == hot space; j = 2 == warm space
|    +--FOR j = 1 TO 2
|    |    n(j, i) = pres(i) * vol(j, i) / (r * t(j))
|    +--NEXT
+--NEXT

     ' record pressures at 90 degree steps
     LOCATE , 1
     fmt$ = "pres 0=##.## MPa    90=##.## MPa    180=##.## MPa    270=##.## MPa"
     PRINT USING fmt$; pres(0); pres(90); pres(180); pres(270)
     PRINT #1, USING fmt$; pres(0); pres(90); pres(180); pres(270)


     '****************************************************************
     ' print max, min and avg pressures
     '****************************************************************
     fmt$ = "pres max=##.## MPa    min=##.## MPa    avg=##.## MPa"
     PRINT USING fmt$; pmax; pmin; pavg
     PRINT #1, USING fmt$; pmax; pmin; pavg


     '****************************************************************
     ' calculate heat in and heat out
     '****************************************************************
     ' step through one complete cycle, with i = 0 to 360 degrees
+--FOR i = 1 TO 360
|    ' trace information
|      LOCATE , 1
|      PRINT "heat in/out"; i;
|
|    ' calculate average pressure between two degree steps
|      apres = (pres(i) + pres(i - 1)) / 2
|
|    ' step through each space
|    +--FOR j = 1 TO 2
|    |
|    |  ' calculate heat based on PdV (average pressure * volume change)
|    |    ht = apres * (vol(j, i) - vol(j, i - 1))
|    |
|    |  ' determine whether heat is flowing in or out
|    |  ' heat in is a positive value; heat out is a negative value
|    |    +--IF ht > 0 THEN
|    |    |    hi(j, i) = ht
|    |    |    ho(j, i) = 0
|    |    +--ELSE
|    |    |    hi(j, i) = 0
|    |    |    ho(j, i) = ht
|    |    +--END IF
|    +--NEXT
+--NEXT


     '****************************************************************
     ' calculate total heat into and out of each space for one complete cycle
```

```
'•••••••••••••••••••••••••••••••••••••••••••••••••••••••••••••••••
  hit(1) = 0
  hit(2) = 0
  hot(1) = 0
  hot(2) = 0
  hn(1) = 0
  hn(2) = 0

  ' step through one complete cycle, with i = 1 to 360 degrees
+--FOR i = 1 TO 360
|   ' step through each space: warm, cold
|  +--FOR j = 1 TO 2
|  |   ' total heat into each space
|  |       hit(j) = hit(j) + hi(j, i)
|  |   ' total heat out of each space
|  |       hot(j) = hot(j) + ho(j, i)
|  |   ' total net heat (into and out of each space)
|  |       hn(j) = hn(j) + hi(j, i) + ho(j, i)
|  +--NEXT
+--NEXT

  ' report and store results in file
  LOCATE , 1

'•••••••••••••••••••••••••••••••••••••••••••••••••••••••••••••••••
  ' calculated results of heat transferred
'•••••••••••••••••••••••••••••••••••••••••••••••••••••••••••••••••
  fmt$ = "temp     1=###### F     2=###### F   "
  PRINT USING fmt$; t(1) * 1.8 - 460; t(2) * 1.8 - 460
  PRINT #1, USING fmt$; t(1) * 1.8 - 460; t(2) * 1.8 - 460
  fmt$ = "heat in  1=######.## J  2=######.## J"
  PRINT USING fmt$; hit(1); hit(2)
  PRINT #1, USING fmt$; hit(1); hit(2)
  fmt$ = "heat out 1=######.## J  2=######.## J"
  PRINT USING fmt$; hot(1); hot(2)
  PRINT #1, USING fmt$; hot(1); hot(2)
  fmt$ = "net heat 1=######.## J  2=######.## J"
  PRINT USING fmt$; hn(1); hn(2)
  PRINT #1, USING fmt$; hn(1); hn(2)

'•••••••••••••••••••••••••••••••••••••••••••••••••••••••••••••••••
  ' calculate mass flow between spaces
'•••••••••••••••••••••••••••••••••••••••••••••••••••••••••••••••••
  n1i = 0
  n1o = 0

  ' step through one complete cycle, with i = 1 to 360 degrees
+--FOR i = 1 TO 360
|   ' determine how much working fluid moved between previous and current steps
|   ' for space 1
|       n1 = n(1, i) - n(1, i - 1)
|   ' determine direction of flow
|  +--IF n1 >= 0 THEN
|  |       n1i = n1i + n1
```

188   Appendix C

```
|  +--ELSE
|  |      n1o = n1o - n1
|  +--END IF
+--NEXT

     ' mass flow average (in and out) for space (to reduce numerical errors)
     n1 = (n1i + n1o) / 2

     ' calculate amount of heat exchanged with regenerator
     ' as gas goes from warm to cold
     hs1 = n1 * 4 * 5.19 * (t(1) - t(2))

     ' print results
     PRINT USING "mass transfer in/out (1-2): #####.####"; n1
     PRINT #1, USING "mass transfer in/out (1-2): #####.####"; n1

     PRINT USING "regenerator storage (1-2): ######.# J"; hs1
     PRINT #1, USING "regenerator storage (1-2): ######.# J"; hs1

     '*************************************************************
     ' calculate work in/out
     '*************************************************************
     wi = 0
     wo = 0
     ' step through one complete cycle, with i = 1 to 360
+--FOR i = 1 TO 360
|    ' calculate work
|      w = (hi(1, i) + hi(2, i) + ho(1, i) + ho(2, i))
|    ' determine direction of flow
|    ' work out of the concept is a positive value;
|    ' work into the concept is a negative value
|  +--IF w < 0 THEN
|  |      wi = wi + w
|  +--ELSE
|  |      wo = wo + w
|  +--END IF
+--NEXT

     ' print out results
     PRINT USING "work out: ######.# J"; wo;
     PRINT USING "     in: ######.# J"; wi;
     PRINT USING "  absorbed: ######.# J"; wo + wi
     PRINT #1, USING "work out: ######.# J"; wo;
     PRINT #1, USING "     in: ######.# J"; wi;
     PRINT #1, USING "  absorbed: ######.# J"; wo + wi


     '*************************************************************
     ' print out details for file
     '*************************************************************
     PRINT #1, "volumes throughout cycle"
+--FOR i = 1 TO 360
|      PRINT #1, USING "###    #####.## cm^3"; i; vol(1, i);
|      PRINT #1, USING "   #####.## cm^3"; vol(2, i)
```

Stirling-Stirling (Displacer-Piston) Heat Pump Programs

```
+--NEXT

   PRINT #1, "pressure throughout cycle"
+--FOR i = 1 TO 360
|     PRINT #1, USING "###   ###.## MPa"; i; pres(i)
+--NEXT

   PRINT #1, "heat transfer in/out of spaces"
+--FOR i = 1 TO 360
|     PRINT #1, USING "###    ####.#### J"; i; hi(1, i) + ho(1, i);
|     PRINT #1, USING "     ####.#### J"; hi(2, i) + ho(2, i)
+--NEXT

   PRINT #1, "mass transfer in/out of spaces"
+--FOR i = 1 TO 360
|     PRINT #1, USING "###"; i;
|  +--FOR j = 1 TO 2
|  |     PRINT #1, USING "    ##.#####"; n(j, i) - n(j, i - 1);
|  +--NEXT
|     PRINT #1,
+--NEXT

   PRINT #1, "work transfer in/out during cycle"
+--FOR i = 1 TO 360
|     w = hi(1, i) + hi(2, i) + ho(1, i) + ho(2, i)
|     PRINT #1, USING "###     ######.#### J"; i; w
+--NEXT

   ' close file
   CLOSE
   END
```

```
'••••••••••••••••••••••••••••••••••••••••••••••••••••••••••••••••••••
' STPD-ENG.BAS - stimulation of a Stirling cycle engine
' for piston-displacer configuration
'••••••••••••••••••••••••••••••••••••••••••••••••••••••••••••••••••••

'••••••••••••••••••••••••••••••••••••••••••••••••••••••••••••••••••••
' set flag for dynamic arrays
' $DYNAMIC

'••••••••••••••••••••••••••••••••••••••••••••••••••••••••••••••••••••
' define variable types
DEFDBL A-H
DEFDBL K-Z
DEFINT I-J

'••••••••••••••••••••••••••••••••••••••••••••••••••••••••••••••••••••
' array definition
'••••••••••••••••••••••••••••••••••••••••••••••••••••••••••••••••••••
' volume for each working space (hot and warm),
' at cycle steps of one degree
DIM vol(2, 360)
' pressure, at one degree steps
DIM pres(360)
' working fluid in each working space, at one degree steps
DIM n(2, 360)
' absolute temperature in each working space
DIM t(2)
' heat into each space, over each degree step
DIM hi(2, 360)
' heat out of each space, over each degree step
DIM ho(2, 360)
' total heat into each space
DIM hit(2)
' total heat out of each space
DIM hot(2)
' net heat transfer of each space
DIM hn(2)

'••••••••••••••••••••••••••••••••••••••••••••••••••••••••••••••••••••
' constants
'••••••••••••••••••••••••••••••••••••••••••••••••••••••••••••••••••••
pi = 3.1415926#
bl$ = SPACE$(40)
' ideal gas constant, cm^3 MPa / K mol
r = 8.3143

'••••••••••••••••••••••••••••••••••••••••••••••••••••••••••••••••••••
' parameter initialization
'••••••••••••••••••••••••••••••••••••••••••••••••••••••••••••••••••••
' initial value for total amount of working fluid, moles
nt = 1
' total volume, cm^3
ttlvol = 1000
' dead space volume fraction
```

Stirling-Stirling (Displacer-Piston) Heat Pump Programs

```
frc.ds = .001
' desired (mean) operating pressure, MPa
aop = 5
' hot temperature, warm temperature, degrees F
th = 1000
tw = 150

'******************************************************************
' computed working values
'******************************************************************
' convert temperatures to kelvin
t(1) = (th + 460) / 1.8
t(2) = (tw + 460) / 1.8
' compute active space volume fraction (no dead space is included)
frc.as = 1 - frc.ds

'******************************************************************
' output operation
'******************************************************************
' record initial values of analysis
CLS
OPEN "stpd-eng.dat" FOR OUTPUT AS 1
PRINT "Stirling engine - piston/displacer -- " + DATE$ + "  | " + TIME$
PRINT #1, "Stirling engine - piston/displacer -- " + DATE$ + "  | " + TIME$

fmt$ = "th=#### F   tw=#### F   dead spc=#.###"
PRINT USING fmt$; th; tw; frc.ds
PRINT #1, USING fmt$; th; tw; frc.ds

'******************************************************************
' calculate volumes for each space for each degree step
'******************************************************************
vol1 = ttlvol / 2 * frc.as
vol2 = vol1
offset = 9999999

' step through one complete cycle, with i = 0 to 360 degrees
+--FOR i = 0 TO 360
|   ' trace information
|       LOCATE , 1
|       PRINT "working on volume"; i;
|
|   ' calculate instantaneous volume for each space,
|   ' accounting for only active space
|       vol(1, i) = vol1 * (1 + SIN((i + 90) * pi / 180) * frc.as) / 2
|       vol(2, i) = vol2 * (1 + SIN((i) * pi / 180) * frc.as) / 2
|
|   ' search for maximum overlap between hot and warm spaces
|   +--IF (vol(2, i) - vol(1, i)) < offset THEN
|   |       offset = vol(2, i) - vol(1, i)
|   |       mi = i
|   +--END IF
|
+--NEXT
```

Appendix C

```
     LOCATE , 1

     ' determine an offsetting value to produce no overlap between
     ' hot and warm spaces
     offset = -offset
     delta = ttlvol * frc.as / (vol1 + offset)

     ' report on adjusted volume parameters
     fmt$ = "ttl=#####.#   v1=#####.#   v2=#####.#   os=#####.#"
     PRINT USING fmt$; ttlvol; vol1; vol2; offset
     PRINT #1, USING fmt$; ttlvol; vol1; vol2; offset

     v1l = 99999
     v2l = 99999
     vmin = 99999
     v1h = 0
     v2h = 0
     vmax = 0

     ' adjust volumes allocated to each space by amount needed to shift
     ' piston away from displacer
+--FOR i = 0 TO 360
|    vol(1, i) = vol(1, i) * delta
|    vol(2, i) = (vol(2, i) + offset) * delta - vol(1, i) + vol2 * frc.ds
|    vol(1, i) = vol(1, i) + vol1 * frc.ds
|
|    ' calculate total instantaneous volume to find total max and min volumes
|    vsum = vol(1, i) + vol(2, i)
|    IF vmax < vsum THEN vmax = vsum
|    IF vmin > vsum THEN vmin = vsum
|    IF vol(1, i) > v1h THEN v1h = vol(1, i)
|    IF vol(1, i) < v1l THEN v1l = vol(1, i)
|    IF vol(2, i) > v2h THEN v2h = vol(2, i)
|    IF vol(2, i) < v2l THEN v2l = vol(2, i)
+--NEXT
     LOCATE , 1

     ' report on volume results
     PRINT USING "vmax=#####.#    vmin=#####.#"; vmax; vmin
     PRINT #1, USING "vmax=#####.#    vmin=#####.#"; vmax; vmin
     PRINT USING "min vol1=#####.#   vol2=#####.#"; v1l; v2l
     PRINT USING "max vol1=#####.#   vol2=#####.#"; v1h; v2h
     PRINT #1, USING "min vol1=#####.#   vol2=#####.#"; v1l; v2l
     PRINT #1, USING "max vol1=#####.#   vol2=#####.#"; v1h; v2h

     fmt$ = "vol 1=####### cc   2=####### cc   ### deg"
     PRINT USING fmt$; vol(1, 0); vol(2, 0); 0
     PRINT USING fmt$; vol(1, 45); vol(2, 45); 45
     PRINT USING fmt$; vol(1, 90); vol(2, 90); 90
     PRINT USING fmt$; vol(1, 135); vol(2, 135); 135
     PRINT USING fmt$; vol(1, 180); vol(2, 180); 180
     PRINT USING fmt$; vol(1, 225); vol(2, 225); 225
     PRINT USING fmt$; vol(1, 270); vol(2, 270); 270
     PRINT USING fmt$; vol(1, 315); vol(2, 315); 315
```

Stirling-Stirling (Displacer-Piston) Heat Pump Programs   193

```
   PRINT #1, USING fmt$; vol(1, 0); vol(2, 0); 0
   PRINT #1, USING fmt$; vol(1, 45); vol(2, 45); 45
   PRINT #1, USING fmt$; vol(1, 90); vol(2, 90); 90
   PRINT #1, USING fmt$; vol(1, 135); vol(2, 135); 135
   PRINT #1, USING fmt$; vol(1, 180); vol(2, 180); 180
   PRINT #1, USING fmt$; vol(1, 225); vol(2, 225); 225
   PRINT #1, USING fmt$; vol(1, 270); vol(2, 270); 270
   PRINT #1, USING fmt$; vol(1, 315); vol(2, 315); 315

   '***************************************************************
   ' calculate pressures at each degree step assuming uniform instantaneous
   ' pressure throughout
   '***************************************************************
   ' step through one complete cycle, with i = 0 to 360 degrees
+--FOR i = 0 TO 360
|     ' trace information
|     LOCATE , 1
|     PRINT "working on pressure"; i;
|     pres(i) = r * nt / (vol(1, i) / t(1) + vol(2, i) / t(2))
+--NEXT

   '***************************************************************
   ' determine max, min, avg pressures
   '***************************************************************
   pmax = pres(1)
   pmin = pres(1)
   pavg = pres(1)

   ' step through cycle
+--FOR i = 2 TO 360
|     IF pres(i) > pmax THEN pmax = pres(i)
|     IF pres(i) < pmin THEN pmin = pres(i)
|     pavg = pavg + pres(i)
+--NEXT
   pavg = pavg / 360

   '***************************************************************
   ' adjust for desired mean operating pressure
   '***************************************************************
   ' determine ratio between desired average operating pressure and
   ' computed average operating pressure
   rto = aop / pavg

   ' adjust total working fluid amount to reflect desired pressure
   ' change from computed value
   nt = nt * rto
   ' calculate new max, min and average values
   pmax = pmax * rto
   pmin = pmin * rto
   pavg = aop

   ' step through one complete cycle, with i = 0 to 360 degrees,
   ' calculating new pressure based on ratio between calculated and
   ' design average
```

```
+--FOR i = 0 TO 360
|    pres(i) = pres(i) * rto
|    ' calculate amount of working fluid in each space for each degree step
|    ' j = 1 == hot space; j = 2 == warm space
|    +--FOR j = 1 TO 2
|    |    n(j, i) = pres(i) * vol(j, i) / (r * t(j))
|    +--NEXT
+--NEXT

     ' record pressures at 90 degree steps
     LOCATE , 1
     fmt$ = "pres 0=##.## MPa    90=##.## MPa    180=##.## MPa    270=##.## MPa"
     PRINT USING fmt$; pres(0); pres(90); pres(180); pres(270)
     PRINT #1, USING fmt$; pres(0); pres(90); pres(180); pres(270)

     '*************************************************************************
     ' print max, min and avg pressures
     '*************************************************************************
     fmt$ = "pres max=##.## MPa    min=##.## MPa    avg=##.## MPa"
     PRINT USING fmt$; pmax; pmin; pavg
     PRINT #1, USING fmt$; pmax; pmin; pavg

     '*************************************************************************
     ' calculate heat in and heat out
     '*************************************************************************
     ' step through one complete cycle, with i = 0 to 360 degrees
+--FOR i = 1 TO 360
|    ' trace information
|      LOCATE , 1
|      PRINT "heat in/out"; i;
|
|    ' calculate average pressure between two degree steps
|      apres = (pres(i) + pres(i - 1)) / 2
|
|    ' step through each space
|    +--FOR j = 1 TO 2
|    |
|    |  ' calculate heat based on Pdv (average pressure * volume change)
|    |    ht = apres * (vol(j, i) - vol(j, i - 1))
|    |
|    |  ' determine whether heat is flowing in or out
|    |  ' heat in is a positive value; heat out is a negative value
|    |    +--IF ht > 0 THEN
|    |    |    hi(j, i) = ht
|    |    |    ho(j, i) = 0
|    |    +--ELSE
|    |    |    hi(j, i) = 0
|    |    |    ho(j, i) = ht
|    |    +--END IF
|    +--NEXT
+--NEXT

     '*************************************************************************
     ' calculate total heat into and out of each space for one complete cycle
```

Stirling-Stirling (Displacer-Piston) Heat Pump Programs 195

```
'*****************************************************
  hit(1) = 0
  hit(2) = 0
  hot(1) = 0
  hot(2) = 0
  hn(1) = 0
  hn(2) = 0

  ' step through one complete cycle, with i = 1 to 360 degrees
+--FOR i = 1 TO 360
|   ' step through each space: hot, warm
|  +--FOR j = 1 TO 2
|  |  ' total heat into each space
|  |     hit(j) = hit(j) + hi(j, i)
|  |  ' total heat out of each space
|  |     hot(j) = hot(j) + ho(j, i)
|  |  ' total net heat (into and out of each space)
|  |     hn(j) = hn(j) + hi(j, i) + ho(j, i)
|  +--NEXT
+--NEXT

  ' report and store results in file
  LOCATE , 1

'*****************************************************
  ' calculated results of heat transferred
'*****************************************************
  fmt$ = "temp     1=###### F     2=###### F    "
  PRINT USING fmt$; t(1) * 1.8 - 460; t(2) * 1.8 - 460
  PRINT #1, USING fmt$; t(1) * 1.8 - 460; t(2) * 1.8 - 460
  fmt$ = "heat in  1=######.## J  2=######.## J"
  PRINT USING fmt$; hit(1); hit(2)
  PRINT #1, USING fmt$; hit(1); hit(2)
  fmt$ = "heat out 1=######.## J  2=######.## J"
  PRINT USING fmt$; hot(1); hot(2)
  PRINT #1, USING fmt$; hot(1); hot(2)
  fmt$ = "net heat 1=######.## J  2=######.## J"
  PRINT USING fmt$; hn(1); hn(2)
  PRINT #1, USING fmt$; hn(1); hn(2)

'*****************************************************
  ' calculate mass flow between spaces
'*****************************************************
  n1i = 0
  n1o = 0

  ' step through one complete cycle, with i = 1 to 360 degrees
+--FOR i = 1 TO 360
|   ' determine how much working fluid moved between previous and current steps
|   ' for space 1 (hot)
|      n1 = n(1, i) - n(1, i - 1)
|   ' determine direction of flow
|  +--IF n1 >= 0 THEN
|  |     n1i = n1i + n1
```

```
|  +--ELSE
|  |     n1o = n1o - n1
|  +--END IF
+--NEXT

    ' mass flow average (in and out) for hot space (to reduce numerical errors)
    n1 = (n1i + n1o) / 2

    ' calculate amount of heat exchanged with regenerator
    ' as gas goes from hot to warm
    hs1 = n1 * 4 * 5.19 * (t(1) - t(2))

    ' print results
    PRINT USING "mass transfer in/out (1-2): #####.####"; n1
    PRINT #1, USING "mass transfer in/out (1-2): #####.####"; n1

    PRINT USING "regenerator storage (1-2): ######.# J"; hs1
    PRINT #1, USING "regenerator storage (1-2): ######.# J"; hs1

    '..........................................................................
    ' calculate work in/out
    '..........................................................................
    wi = 0
    wo = 0
    ' step through one complete cycle, with i = 1 to 360 degrees
+--FOR i = 1 TO 360
|   ' calculate work
|       w = (hi(1, i) + hi(2, i) + ho(1, i) + ho(2, i))
|   ' determine direction of flow
|   ' work out of the concept is a positive value;
|   ' work into the concept is a negative value
|   +--IF w < 0 THEN
|   |      wi = wi + w
|   +--ELSE
|   |      wo = wo + w
|   +--END IF
+--NEXT

    ' print out results
    PRINT USING "work out: ######.# J"; wo;
    PRINT USING "      in: ######.# J"; wi;
    PRINT USING "     net: ######.# J"; wo + wi
    PRINT #1, USING "work out: ######.# J"; wo;
    PRINT #1, USING "      in: ######.# J"; wi;
    PRINT #1, USING "     net: ######.# J"; wo + wi

    '..........................................................................
    ' print out details for file
    '..........................................................................
    PRINT #1, "volumes throughout cycle"
+--FOR i = 1 TO 360
|       PRINT #1, USING "###   #####.## cm^3"; i; vol(1, i);
|       PRINT #1, USING "    #####.## cm^3"; vol(2, i)
+--NEXT
```

```
    PRINT #1, "pressure throughout cycle"
+--FOR i = 1 TO 360
|     PRINT #1, USING "###    ###.## MPa"; i; pres(i)
+--NEXT

    PRINT #1, "heat transfer in/out of spaces"
+--FOR i = 1 TO 360
|     PRINT #1, USING "###    ####.#### J"; i; hi(1, i) + ho(1, i);
|     PRINT #1, USING "    ####.#### J"; hi(2, i) + ho(2, i)
+--NEXT

    PRINT #1, "mass transfer in/out of spaces"
+--FOR i = 1 TO 360
|     PRINT #1, USING "###"; i;
|  +--FOR j = 1 TO 2
|  |     PRINT #1, USING "    ##.#####"; n(j, i) - n(j, i - 1);
|  +--NEXT
|     PRINT #1,
+--NEXT

    PRINT #1, "work transfer in/out during cycle"
+--FOR i = 1 TO 360
|     w = -(hi(1, i) + hi(2, i) + ho(1, i) + ho(2, i))
|     PRINT #1, USING "###    ######.#### J"; i; w
+--NEXT

    ' close file
    CLOSE
    END
```

Appendix
D
Mechanical-Compression Vuilleumier Heat Pump Program

The MCVUILL.BAS program, written in QuickBASIC version 4.0 (a trademark of Microsoft Corporation), simulates the ideal, isothermal operation of the balanced-compounded embodiment of a mechanical-compression Vuilleumier heat pump.

The program works by determining the volume variations for each of the three spaces for one degree increments through one complete cycle of operation. It then determines the pressure variations through one cycle. Using the ratio of the computed average value to the design value for the mean operating pressure, the program adjusts the pressures for each cycle step to meet the design mean operating pressure. It also adjusts the amount of working fluid in the balanced-compounded Vuilleumier to correspond to the thermodynamic conditions in each space.

Before executing, values of the following parameters must be specified within the program:

Parameter name	Parameter definition	Units
nt	Starting value for total quantity of working fluid	gram-moles
ttlvol	Total volume of all working spaces	cm^3
v3.vt	Ratio between the cold-space volume (volume 3) and the total volume	None
frc.ds	Fraction of the total volume designated as dead-space volume	None
aop	Average (mean) operating pressure	MPa
th	High temperature (hot space)	°F
ti	Intermediate temperature (warm space)	°F
tc	Low temperature (cold space)	°F

The above parameters are contained in the section marked *parameter initialization*. The program calculates the heat exchanged with the environment, the mass transfer between spaces, the heat storage capacity of the regenerator assuming 100% effective operation, and the work storage profile for one cycle.

To find the solution, the user must iteratively adjust one parameter. This parameter, v3.vt, the volume ratio between the cold-space volume and the total volume, needs to be adjusted to reflect whether net work is produced or consumed. If the heat in does not equal the heat out for a particular cold-space volume fraction, the balanced-compounded Vuilleumier either requires work input to operate or delivers work to the outside in addition to providing heat pumping. When the heat in equals the heat out among the three spaces, there is only heat pumping; no net work is produced or consumed over the cycle. If an application either uses work or produces work in addition to heat pumping, the user must adjust v3.vt until the program reports values for heat in and heat out that correspond to the work. If the application only provides heat pumping, the user must adjust v3.vt until the program reports equal values for heat in and heat out.

When the correct cold-space volume fraction has been determined, the machine size required to meet the design capacity is calculated by multiplying the working volume used by the ratio between the design value of the cooling capacity and calculated capacity.

During execution, the program displays summary values for the simulation. It also displays a counter indicating the progress of the calculation. The summary results consist of:

- Maximum and minimum total working volume, cm^3
- Volume of each space at the 0, 90, 180, and 270° cycle steps, cm^3
- Pressure at the 0, 90, 180, and 270° cycle steps, MPa
- The maximum, minimum, and average pressures, MPa
- Heat into each space, J
- Heat out of each space, J
- Net heat transfer between the space and the environment, J
- Quantity of working fluid that is transferred between the hot and warm spaces (1 and 2), gram-moles
- Quantity of working fluid that is transferred between the warm and cold spaces (2 and 3), gram-moles
- Regenerator storage capacity needed to achieve 100% effectiveness

(equivalent to the assumption of no heat transfer between spaces due to mass transfer), J

- Work that must be transferred between various parts of the operating cycle, requiring some form of storage external to the working fluid (such as a moving mass), J

In addition to displaying these results on the monitor, the program stores these values in an ASCII text file named MCVUILL.DAT. Values for volumes, pressures, heat transfer between each space and the environment, mass transfer between spaces, and work transfer between the cycle and the environment are stored in 1° steps for the last complete operating cycle in MCVUILL.DAT.

Appendix D

```
'*****************************************************************
' MCVUILL.BAS - simulation of a mechanical compression Vuilleumier
' in balanced-compounded Vuilleumier configuration
'*****************************************************************


'*****************************************************************
' set flag for dynamic arrays
' $DYNAMIC

'*****************************************************************
' define variable types
DEFDBL A-H
DEFDBL K-Z
DEFINT I-J

'*****************************************************************
' array definition
'*****************************************************************
' volume for each working space (hot, warm and cold),
' at cycle steps of one degree
DIM vol(3, 360)
' pressure, at one degree steps
DIM pres(360)
' working fluid in each working space, at one degree steps
DIM n(3, 360)
' absolute temperature in each of three working spaces
DIM t(3)
' heat into each space, over each degree step
DIM hi(3, 360)
' heat out of each space, over each degree step
DIM ho(3, 360)
' total heat into each space
DIM hit(3)
' total heat out of each space
DIM hot(3)
' net heat transfer of each space
DIM hn(3)

'*****************************************************************
' constants
'*****************************************************************
pi = 3.1415926#
bl$ = SPACES$(40)
' ideal gas constant, cm^3 MPa / K mol
r = 8.3143

'*****************************************************************
' parameter initialization
'*****************************************************************
' initial value for total amount of working fluid, moles
nt = 1
' total volume, cm^3
ttlvol = 1000
' volume ratio between cold space (volume 3) and total volume
```

Mechanical-Compression Vuilleumier Heat Pump Program

```
     v3.vt = .7082
    ' dead space volume fraction
     frc.ds = .001
    ' desired average (mean) operating pressure, MPa
     aop = 5
    ' hot temperature, intermediate temperature, cold temperature, degrees F
     th = 1000
     ti = 150
     tc = 32

    '*********************************************************************
    ' computed working values
    '*********************************************************************
    ' convert temperatures to kelvin
     t(1) = (th + 460) / 1.8
     t(2) = (ti + 460) / 1.8
     t(3) = (tc + 460) / 1.8
    ' compute active space volume fraction (no dead space is included)
     frc.as = 1 - frc.ds
    ' maximum value for volume 3 (cold)
     vol3 = ttlvol * v3.vt
    ' maximum value for volume 1 (hot)
     vol1 = (ttlvol - vol3)
    ' maximum value for volume 2 (intermediate/warm)
     vol2 = vol1

    '*********************************************************************
    ' output operation
    '*********************************************************************
    ' record initial values of analysis
     CLS
     OPEN "mcvuill.dat" FOR OUTPUT AS 1
     PRINT "Mechanical Compression Vuilleumier -- " + DATE$ + " | " + TIME$
     PRINT #1, "Mechanical Compression Vuilleumier -- " + DATE$ + " | " + TIME$

     fmt$ = "Total #####.# cm^3,    v3 = #.#### (frc)"
     PRINT USING fmt$; ttlvol; v3.vt
     PRINT #1, USING fmt$; ttlvol; v3.vt

     fmt$ = "th = #### F    ti = #### F    tc = #### F    dead spc = #.###"
     PRINT USING fmt$; th; ti; tc; frc.ds
     PRINT #1, USING fmt$; th; ti; tc; frc.ds

    '*********************************************************************
    ' set up initial placeholders for maximum and minimum total volumes
    '*********************************************************************
     vmax = -100000
     vmin = 100000

    '*********************************************************************
    ' calculate volumes for each space for each degree step
    '*********************************************************************
    ' step through one complete cycle, with i = 0 to 360 degrees
+--  FOR i = 0 TO 360
```

```
|   ' trace information
|     LOCATE , 1
|     PRINT "working on volume"; i;
|
|   ' calculate instantaneous volume for each space,
|   ' accounting for only active space
|     vol(1, i) = vol1 * (1 + SIN(i * pi / 180) * frc.as) / 2
|     vol(2, i) = vol2 * (1 + SIN((i - 90) * pi / 180) * frc.as) / 2
|     vol(3, i) = vol3 * (1 + SIN(i * pi / 180) * frc.as) / 2
|
|   ' calculate total instantaneous volume to find total max and min volumes
|     vsum = vol(1, i) + vol(2, i) + vol(3, i)
|     IF vmax < vsum THEN vmax = vsum
|     IF vmin > vsum THEN vmin = vsum
+---NEXT
    LOCATE , 1

    ' report on volume results
    PRINT USING "vmax = #####.##"; vmax;
    PRINT USING "     vmin = #####.##"; vmin
    PRINT #1, USING "vmax = #####.##"; vmax;
    PRINT #1, USING "     vmin = #####.##"; vmin

    fmt$ = "vol 1 = ####### cc   2 = ####### cc   3 = ####### cc    ### deg"
    PRINT USING fmt$; vol(1, 0); vol(2, 0); vol(3, 0); 0
    PRINT USING fmt$; vol(1, 90); vol(2, 90); vol(3, 90); 90
    PRINT USING fmt$; vol(1, 180); vol(2, 180); vol(3, 180); 180
    PRINT USING fmt$; vol(1, 270); vol(2, 270); vol(3, 270); 270
    PRINT USING fmt$; vol(1, 360); vol(2, 360); vol(3, 360); 360
    PRINT #1, USING fmt$; vol(1, 0); vol(2, 0); vol(3, 0); 0
    PRINT #1, USING fmt$; vol(1, 90); vol(2, 90); vol(3, 90); 90
    PRINT #1, USING fmt$; vol(1, 180); vol(2, 180); vol(3, 180); 180
    PRINT #1, USING fmt$; vol(1, 270); vol(2, 270); vol(3, 270); 270
    PRINT #1, USING fmt$; vol(1, 360); vol(2, 360); vol(3, 360); 360

    '********************************************************************
    ' calculate pressures at each degree step assuming uniform instantaneous
    ' pressure throughout
    '********************************************************************
    ' step through one complete cycle, with i = 0 to 360 degrees
+---FOR i = 0 TO 360
|   ' trace information
|     LOCATE , 1
|     PRINT "working on pressure"; i;
|
|     pres(i) = r * nt / (vol(1, i) / t(1) + vol(2, i) / t(2) + vol(3, i) / t(3))
|
+---NEXT

    '********************************************************************
    ' determine max, min, avg pressures
    '********************************************************************
    pmax = pres(1)
    pmin = pres(1)
```

```
      pavg = pres(1)
     ' step through cycle
+--FOR i = 2 TO 360
|       IF pres(i) > pmax THEN pmax = pres(i)
|       IF pres(i) < pmin THEN pmin = pres(i)
|       pavg = pavg + pres(i)
+--NEXT
      pavg = pavg / 360

     '*********************************************************
     ' adjust for desired mean operating pressure
     '*********************************************************
     ' determine ratio between desired average operating pressure and
     ' computed average operating pressure
      rto = aop / pavg

     ' adjust total working fluid amount to reflect desired pressure
     ' change from computed value
      nt = nt * rto
     ' calculate new max, min and average values
      pmax = pmax * rto
      pmin = pmin * rto
      pavg = aop

     ' step through one complete cycle, with i = 0 to 360 degrees,
     ' calculating new pressure for each step based on change in
     ' amount of working fluid in moles
+--FOR i = 0 TO 360
|       pres(i) = pres(i) * rto
|     ' calculate amount of working fluid in each space for each degree step
|     ' j = 1 == hot space; j = 2 == warm space; j = 3 == cold space
|     +--FOR j = 1 TO 3
|     |    n(j, i) = pres(i) * vol(j, i) / (r * t(j))
|     +--NEXT
+--NEXT

     ' record pressures at 90 degree steps
      LOCATE , 1
      fmt$ = "pres 0 = ##.## MPa     90 = ##.## MPa     180 = ##.## MPa     270 = ##.## MPa"
      PRINT USING fmt$; pres(0); pres(90); pres(180); pres(270)
      PRINT #1, USING fmt$; pres(0); pres(90); pres(180); pres(270)

     '*********************************************************
     ' print max, min and avg pressures
     '*********************************************************
      fmt$ = "pres max = ##.## MPa     min = ##.## MPa     avg = ##.## MPa"
      PRINT USING fmt$; pmax; pmin; pavg
      PRINT #1, USING fmt$; pmax; pmin; pavg

     '*********************************************************
     ' calculate heat in and heat out
     '*********************************************************
     ' step through one complete cycle, with i = 0 to 360 degrees
```

```
+--FOR i = 1 TO 360
|  ' trace information
|     LOCATE , 1
|     PRINT "heat in/out"; i;
|
|  ' calculate average pressure between two degree steps
|     apres = (pres(i) + pres(i - 1)) / 2
|
|  ' step through each space
|  +--FOR j = 1 TO 3
|  |
|  ' calculate heat based on PdV (average pressure * volume change)
|  |    ht = apres * (vol(j, i) - vol(j, i - 1))
|  |
|  ' determine whether heat is flowing in or out
|  ' heat in is a positive value; heat out is a negative value
|  |  +--IF ht > 0 THEN
|  |  |     hi(j, i) = ht
|  |  |     ho(j, i) = 0
|  |  +--ELSE
|  |  |     hi(j, i) = 0
|  |  |     ho(j, i) = ht
|  |  +--END IF
|  |
|  +--NEXT
|
+--NEXT

   '•••••••••••••••••••••••••••••••••••••••••••••••••••••••••••••••••••
   ' calculate total heat into and out of each space for one complete cycle
   '•••••••••••••••••••••••••••••••••••••••••••••••••••••••••••••••••••
   hit(1) = 0
   hit(2) = 0
   hit(3) = 0
   hot(1) = 0
   hot(2) = 0
   hot(3) = 0
   hn(1) = 0
   hn(2) = 0
   hn(3) = 0

   ' step through one complete cycle, with i = 1 to 360 degrees
+--FOR i = 1 TO 360
|  ' step through each space, hot, warm, cold
|  +--FOR j = 1 TO 3
|  ' total heat into each space
|  |    hit(j) = hit(j) + hi(j, i)
|  ' total heat out of each space
|  |    hot(j) = hot(j) + ho(j, i)
|  ' total net heat (into and out of each space)
|  |    hn(j) = hn(j) + hi(j, i) + ho(j, i)
|  +--NEXT
+--NEXT
```

Mechanical-Compression Vuilleumier Heat Pump Program

```
     ' report and store results in file
     LOCATE , 1
     fmt$ = "heat 1 = ######.## J  2 = ######.## J  3 = ######.## J    ### deg"
     ' cycle through 90 degree steps
+--FOR i = 90 TO 360 STEP 90
|      PRINT USING fmt$; hi(1, i) + ho(1, i); hi(2, i) + ho(2, i);
|         hi(3, i) + ho(3, i); i
|      PRINT #1, USING fmt$; hi(1, i) + ho(1, i); hi(2, i) + ho(2, i);
|         hi(3, i) + ho(3, i); i
+--NEXT

     '•••••••••••••••••••••••••••••••••••••••••••••••••••••••••••••••••
     ' calculated results of heat transferred
     '•••••••••••••••••••••••••••••••••••••••••••••••••••••••••••••••••
     fmt$ = "heat in  1 = ######.## J  2 = ######.## J  3 = ######.## J"
     PRINT USING fmt$; hit(1); hit(2); hit(3)
     PRINT #1, USING fmt$; hit(1); hit(2); hit(3)
     fmt$ = "heat out 1 = ######.## J  2 = ######.## J  3 = ######.## J"
     PRINT USING fmt$; hot(1); hot(2); hot(3)
     PRINT #1, USING fmt$; hot(1); hot(2); hot(3)
     fmt$ = "net heat 1 = ######.## J  2 = ######.## J  3 = ######.## J"
     PRINT USING fmt$; hn(1); hn(2); hn(3)
     PRINT #1, USING fmt$; hn(1); hn(2); hn(3)

     PRINT USING "net: ######.## J"; hn(1) + hn(2) + hn(3)
     PRINT #1, USING "net: ######.## J"; hn(1) + hn(2) + hn(3)

     '•••••••••••••••••••••••••••••••••••••••••••••••••••••••••••••••••
     ' calculate mass flow between spaces
     '•••••••••••••••••••••••••••••••••••••••••••••••••••••••••••••••••
     n1i = 0
     n1o = 0
     n3i = 0
     n3o = 0

     ' step through one complete cycle, with i = 1 to 360 degrees
+--FOR i = 1 TO 360
|    ' determine how much working fluid moved between previous and current steps
|    ' for space 1 (hot)
|        n1 = n(1, i) - n(1, i - 1)
|    ' determine direction of flow
|    +--IF n1 >= 0 THEN
|    |      n1i = n1i + n1
|    +--ELSE
|    |      n1o = n1o - n1
|    +--END IF
|
|    ' determine how much working fluid moved between previous and current steps
|    ' for space 3 (cold)
|        n3 = n(3, i) - n(3, i - 1)
|    ' determine direction of flow
|    +--IF n3 >= 0 THEN
|    |      n3i = n3i + n3
|    +--ELSE
```

```
|  |       n3o = n3o - n3
|  +--END IF
+--NEXT

    ' mass flow average (in and out) for hot space (to reduce numerical errors)
    n1 = (n1i + n1o) / 2
    ' mass flow average (in and out) for cold space (to reduce numerical errors)
    n3 = (n3i + n3o) / 2
    ' calculate amount of heat exchanged as gas goes from hot to warm
    hs1 = n1 * 4 * 5.19 * (1000 - 150) * 5 / 9
    ' calculate amount of heat exchanged as gas goes from warm to cold
    hs3 = n3 * 4 * 5.19 * (150 - 32) * 5 / 9

    ' print results
    PRINT USING "mass transfer in/out (1-2): #####.####"; n1;
    PRINT USING "     (2-3): #####.####"; n3
    PRINT #1, USING "mass transfer in/out (1-2): #####.####"; n1;
    PRINT #1, USING "     (2-3): #####.####"; n3

    PRINT USING "regenerator storage (1-2): ######.# J"; hs1;
    PRINT USING "     (2-3): ######.# J"; hs3
    PRINT #1, USING "regenerator storage (1-2): ######.# J"; hs1;
    PRINT #1, USING "     (2-3): ######.# J"; hs3

    '***********************************************************************
    ' calculate work in/out
    '***********************************************************************
    wi = 0
    wo = 0
    ' step through one complete cycle, with i = 1 to 360 degrees
+--FOR i = 1 TO 360
|   ' calculate work
|       w = (hi(1, i) + hi(2, i) + hi(3, i) + ho(1, i) + ho(2, i) + ho(3, i))
|   ' determine direction of flow
|   ' work out of the concept is a positive value;
|   ' work into the concept is a negative value
|   +--IF w < 0 THEN
|   |     wi = wi + w
|   +--ELSE
|   |     wo = wo + w
|   +--END IF
+--NEXT

    ' work average (in and out) (to reduce numerical errors)
    w = (-wi + wo) / 2

    ' print results
    PRINT USING "work storage: ######.# J"; w
    PRINT #1, USING "work storage: ######.# J"; w

    '***********************************************************************
    ' print out details for file
    '***********************************************************************
    PRINT #1, "volumes throughout cycle"
```

```
+--FOR i = 1 TO 360
|    PRINT #1, USING "###    #####.## cm^3"; i; vol(1, i);
|    PRINT #1, USING "       #####.## cm^3"; vol(2, i);
|    PRINT #1, USING "       #####.## cm^3"; vol(3, i)
+--NEXT

   PRINT #1, "pressure throughout cycle"
+--FOR i = 1 TO 360
|    PRINT #1, USING "###    ###.## MPa"; i; pres(i)
+--NEXT

   PRINT #1, "heat transfer in/out of spaces"
+--FOR i = 1 TO 360
|    PRINT #1, USING "###    ####.#### J"; i; hi(1, i) + ho(1, i);
|    PRINT #1, USING "       ####.#### J"; hi(2, i) + ho(2, i);
|    PRINT #1, USING "       ####.#### J"; hi(3, i) + ho(3, i)
+--NEXT

   PRINT #1, "mass transfer in/out of spaces"
   ' step through one complete cycle, with i = 1 to 360 degrees
+--FOR i = 1 TO 360
|    PRINT #1, USING "###"; i;
|    ' step through each space
|  +--FOR j = 1 TO 3
|  |    PRINT #1, USING "   ##.#####"; n(j, i) - n(j, i - 1);
|  +--NEXT
|    PRINT #1,
+--NEXT

   PRINT #1, "work transfer in/out during cycle"
   ' step through one complete cycle, with i = 1 to 360 degrees
+--FOR i = 1 TO 360
|    w = -(hi(1, i) + hi(2, i) + hi(3, i) + ho(1, i) + ho(2, i) + ho(3, i))
|    PRINT #1, USING "###    ######.#### J"; i; w
+--NEXT

   ' close file
   CLOSE
   END
```

Appendix

Ericsson-Ericsson Heat Pump Program

The ERICSSON.BAS program, written in QuickBASIC version 4.0 (a trademark of Microsoft Corporation), simulates the ideal, isothermal operation of the Ericsson-Ericsson heat pump (which is thermodynamically similar to the mechanical-compression Vuilleumier heat pump).

The program calculates the volume variations for each of the three spaces for one degree increments through one complete cycle of operation. It then determines the pressure variations through one cycle. Using the ratio of the computed average value to the design value for the mean operating pressure, the program adjusts the pressures for each cycle step to meet the design mean operating pressure. It also adjusts the amount of working fluid in the heat pump to correspond to the thermodynamic condition in each space.

Before executing, values of the following parameters must be specified within the program:

Parameter name	Parameter definition	Units
nt	Starting value for total quantity of working fluid	gram-moles
ttlvol	Total volume for all working spaces	cm^3
vp.vt	Ratio of the warm-space volume (both between the displacers and between the flywheel pistons) to the total volume	None
vc.vh	Ratio of the cold-space volume to the hot-space volume	None
frc.ds	Fraction of the total volume designated as dead-space volume	None
aop	Average (mean) operating pressure	MPa
th	High temperature (hot space)	°F
ti	Intermediate temperature (warm space)	°F
tc	Low temperature (cold space)	°F

The above parameters are contained in the section marked *parameter initialization*. The program calculates the heat exchanged with the environment, the mass transfer between spaces, the heat storage capacity of the regenerator assuming 100% effective operation, and the work storage profile for one cycle.

To specify a particular design, the user must adjust two parameters. The first is vc.vh, the ratio of the cold-space volume to the hot-space volume. This parameter needs to be adjusted to reflect whether net work is produced or consumed. If the heat in does not equal the heat out for a particular cold-space volume to hot-space volume ratio, the Ericsson-Ericsson either requires work input to operate or delivers work in addition to providing heat pumping. When the heat in equals the heat out among the three spaces, there is only heat pumping; no net work is produced or consumed over the cycle. If an application either uses work or produces work in addition to heat pumping, the user must adjust vc.vh until the program reports values for heat in and heat out that correspond to the work. If the application only provides heat pumping, the user must adjust vc.vh until the program reports equal values for heat in and heat out.

The second parameter that needs to be adjusted is vp.vt, the ratio of warm-space volume (both between displacers and between flywheel pistons) to total volume. The degree of mechanical compression is strongly dependent on this term, which affects the total capacity a given working volume can deliver. This term's value should be adjusted to produce the highest specific capacity. Although it is possible to use a different vp.vt value, the value producing the highest capacity for a given volume will result in the smallest working volume for the design capacity. This will minimize the heat transfer loads placed on heat exchangers and regenerators.

When the correct vc.vh and vp.vt values have been determined, the machine size required to meet the design capacity is calculated by multiplying the total volume used by the ratio between the design value of the cooling capacity and calculated capacity.

During execution, the program displays summary values for the simulation. It also displays a counter indicating the progress of the calculations. The summary results consist of:

- Maximum and minimum total working volume, cm^3
- Volume of each space at the 0, 90, 180, and 270° cycle steps, cm^3
- Pressure at the 0, 90, 180, and 270° cycle steps, MPa
- Maximum, minimum, and average pressures, MPa
- Heat into each space, J

- Heat out of each space, J
- Net heat transfer between each space and the environment, J
- Quantity of working fluid that is transferred between the hot and warm spaces (1 and 2), gram-moles
- Quantity of working fluid that is transferred between the warm and cold spaces (2 and 3), gram-moles
- Regenerator storage capacity needed to achieve 100% effectiveness (equivalent to the assumption of no heat transfer between spaces due to mass transfer), J
- Work that must be transferred between various parts of the operating cycle, requiring some form of storage external to the working fluid (such as a moving mass), J

In addition to displaying these results on the monitor, the program stores these values in an ASCII text file named ERICSSON.DAT. Values for volumes, pressures, heat transfer between each space and the environment, mass transfer between spaces, and work transfer between the cycle and the environment are stored in 1° steps for the last complete operating cycle in ERICSSON.DAT.

214   Appendix E

```
'•••••••••••••••••••••••••••••••••••••••••••••••••••••••••••••••••••••
' ERICSSON.BAS - simulation of a mechanical compression Vuilleumier
' in an Ericsson - Ericsson configuration
'•••••••••••••••••••••••••••••••••••••••••••••••••••••••••••••••••••••

'•••••••••••••••••••••••••••••••••••••••••••••••••••••••••••••••••••••
' set flag for dynamic arrays
' $DYNAMIC

'•••••••••••••••••••••••••••••••••••••••••••••••••••••••••••••••••••••
' define variable types
DEFDBL A-H
DEFDBL K-Z
DEFINT I-J

'•••••••••••••••••••••••••••••••••••••••••••••••••••••••••••••••••••••
' array definition
'•••••••••••••••••••••••••••••••••••••••••••••••••••••••••••••••••••••
' volume for each working space (hot, warm and cold),
' at cycle steps of one degree
DIM vol(3, 360)
' pressure, at one degree steps
DIM pres(360)
' working fluid in each working space, at one degree steps
DIM n(3, 360)
' absolute temperature in each of three working spaces
DIM t(3)
' heat into each space, over each degree step
DIM hi(3, 360)
' heat out of each space, over each degree step
DIM ho(3, 360)
' total heat into each space
DIM hit(3)
' total heat out of each space
DIM hot(3)
' net heat transfer of each space
DIM hn(3)

'•••••••••••••••••••••••••••••••••••••••••••••••••••••••••••••••••••••
' constants
'•••••••••••••••••••••••••••••••••••••••••••••••••••••••••••••••••••••
pi = 3.1415926#
bl$ = SPACE$(40)
' ideal gas constant, cm^3 MPa / K mol
r = 8.3143

'•••••••••••••••••••••••••••••••••••••••••••••••••••••••••••••••••••••
' parameter initialization
'•••••••••••••••••••••••••••••••••••••••••••••••••••••••••••••••••••••
' initial value for total amount of working fluid, moles
nt = 1
' total volume, cm^3
ttlvol = 1000
' volume fraction of space between flywheel pistons to total volume
```

```
vp.vt = .62
' volume ratio between cold space (volume 3) and hot space (volume 1)
vc.vh = 2.4274
' calculated hot space volume / cold space volume
vh.vc = 1 / vc.vh
' dead space volume fraction
frc.ds = .001
' desired average (mean) operating pressure, MPa
aop = 5
' hot temperature, warm temperature, cold temperature, degrees F
th = 1000
ti = 150
tc = 32

'........................................................................
' computed working values
'........................................................................
' convert temperatures to kelvin
t(1) = (th + 460) / 1.8
t(2) = (ti + 460) / 1.8
t(3) = (tc + 460) / 1.8
' compute active space volume fraction (no dead space is included)
frc.as = 1 - frc.ds
' maximum volume for warm space and between flywheel pistons
vp = ttlvol * vp.vt
' maximum volume for cold space
vol3 = (ttlvol - vp) / (vh.vc + 1)
' maximum volume for hot space
vol1 = vol3 * vh.vc


'........................................................................
' output operation
'........................................................................
' record initial values of analysis
CLS
OPEN "ericsson.dat" FOR OUTPUT AS 1
PRINT "Ericsson/Ericsson (mcV) Heat Pump -- " + DATE$ + " | " + TIME$
PRINT #1, "Ericsson/Ericsson (mcV) Heat Pump -- " + DATE$ + " | " + TIME$

fmt$ = "Total #####.# cm^3,    vp=#.#### (frc),    vc/vh=#.#### (frc)"
PRINT USING fmt$; ttlvol; vp.vt; vc.vh
PRINT #1, USING fmt$; ttlvol; vp.vt; vc.vh

fmt$ = "th=#### F    ti=#### F    tc=#### F    dead spc=#.###"
PRINT USING fmt$; th; ti; tc; frc.ds
PRINT #1, USING fmt$; th; ti; tc; frc.ds

'........................................................................
' set up initial placeholders for maximum and minimum total volumes
'........................................................................
vmax = -100000
vmin = 100000


'........................................................................
```

Appendix E

```
' calculate volumes for each space for each degree step
'••••••••••••••••••••••••••••••••••••••••••••••••••••••••••••••••••••••••
    ' step through one complete cycle, with i = 0 to 360 degrees
+--FOR i = 0 TO 360
|   ' trace information
|     LOCATE , 1
|     PRINT "working on volume"; i;
|
|   ' calculate instantaneous volume for each space,
|   ' accounting for only active space
|     vol(1, i) = vol1 * (1 + SIN(i * pi / 180) * frc.as) / 2
|     vol(3, i) = vol3 * (1 + SIN(i * pi / 180) * frc.as) / 2
|     vol(2, i) = vol1 - vol(1, i) + vol3 - vol(3, i)
|     vol(2, i) = vol(2, i) + vp * (1 + SIN((i - 90) * pi / 180) * frc.as) / 2
|
|   ' calculate total instantaneous volume to find total max and min volumes
|     vsum = vol(1, i) + vol(2, i) + vol(3, i)
|     IF vmax < vsum THEN vmax = vsum
|     IF vmin > vsum THEN vmin = vsum
+--NEXT
    LOCATE , 1

    ' report on volume results
    PRINT USING "vmax = #####.##"; vmax;
    PRINT USING "      vmin = #####.##"; vmin
    PRINT #1, USING "vmax = #####.##"; vmax;
    PRINT #1, USING "      vmin = #####.##"; vmin

    fmt$ = "vol 1=####### cc   2=####### cc   3=####### cc   ### deg"
    PRINT USING fmt$; vol(1, 0); vol(2, 0); vol(3, 0); 0
    PRINT USING fmt$; vol(1, 90); vol(2, 90); vol(3, 90); 90
    PRINT USING fmt$; vol(1, 180); vol(2, 180); vol(3, 180); 180
    PRINT USING fmt$; vol(1, 270); vol(2, 270); vol(3, 270); 270
    PRINT USING fmt$; vol(1, 360); vol(2, 360); vol(3, 360); 360
    PRINT #1, USING fmt$; vol(1, 0); vol(2, 0); vol(3, 0); 0
    PRINT #1, USING fmt$; vol(1, 90); vol(2, 90); vol(3, 90); 90
    PRINT #1, USING fmt$; vol(1, 180); vol(2, 180); vol(3, 180); 180
    PRINT #1, USING fmt$; vol(1, 270); vol(2, 270); vol(3, 270); 270
    PRINT #1, USING fmt$; vol(1, 360); vol(2, 360); vol(3, 360); 360

'••••••••••••••••••••••••••••••••••••••••••••••••••••••••••••••••••••••••
    ' calculate pressures at each degree step assuming uniform instantaneous
    ' pressure throughout
'••••••••••••••••••••••••••••••••••••••••••••••••••••••••••••••••••••••••
    ' step through one complete cycle, with i = 0 to 360 degrees
+--FOR i = 0 TO 360
|   ' trace information
|     LOCATE , 1
|     PRINT "working on pressure"; i;
|
|     pres(i) = r * nt / (vol(1, i) / t(1) + vol(2, i) / t(2) + vol(3, i) / t(3))
|
+--NEXT
```

```
'********************************************************
' determine max, min, avg pressures
'********************************************************
  pmax = pres(1)
  pmin = pres(1)
  pavg = pres(1)
  ' step through cycle
+--FOR i = 2 TO 360
|     IF pres(i) > pmax THEN pmax = pres(i)
|     IF pres(i) < pmin THEN pmin = pres(i)
|     pavg = pavg + pres(i)
+--NEXT
  pavg = pavg / 360

'********************************************************
' adjust for desired mean operating pressure
'********************************************************
  ' determine ratio between desired average operating pressure and
  ' computed average operating pressure
  rto = aop / pavg

  ' adjust total working fluid amount to reflect desired pressure
  ' change from computed value
  nt = nt * rto
  ' calculate new max, min and average values
  pmax = pmax * rto
  pmin = pmin * rto
  pavg = aop

  ' step through one complete cycle, with i = 0 to 360 degrees,
  ' calculating new pressure for each step based on change in
  ' amount of working fluid in moles
+--FOR i = 0 TO 360
|     pres(i) = pres(i) * rto
|     ' calculate amount of working fluid in each space for each degree step
|     ' j = 1 == hot space; j = 2 == warm space; j = 3 == cold space
|   +--FOR j = 1 TO 3
|   |    n(j, i) = pres(i) * vol(j, i) / (r * t(j))
|   +--NEXT
+--NEXT

  ' record pressures at 90 degree steps
  LOCATE , 1
  fmt$ = "pres 0=##.## MPa    90=##.## MPa    180=##.## MPa    270=##.## MPa"
  PRINT USING fmt$; pres(0); pres(90); pres(180); pres(270)
  PRINT #1, USING fmt$; pres(0); pres(90); pres(180); pres(270)

'********************************************************
' print max, min and avg pressures
'********************************************************
  fmt$ = "pres max=##.## MPa    min=##.## MPa    avg=##.## MPa"
  PRINT USING fmt$; pmax; pmin; pavg
  PRINT #1, USING fmt$; pmax; pmin; pavg
```

Appendix E

```
'**************************************************************
' calculate heat in and heat out
'**************************************************************
    ' step through one complete cycle, with i = 0 to 360 degrees
+--FOR i = 1 TO 360
|   ' trace information
|     LOCATE , 1
|     PRINT "heat in/out"; i;
|
|   ' calculate average pressure between two degree steps
|     apres = (pres(i) + pres(i - 1)) / 2
|
|   ' step through each space
|   +--FOR j = 1 TO 3
|   |
|   | ' calculate heat based on PdV (average pressure * volume change)
|   |     ht = apres * (vol(j, i) - vol(j, i - 1))
|   |
|   | 'determine whether heat is flowing in or out
|   | ' heat in is a positive value; heat out is a negative value
|   | +--IF ht > 0 THEN
|   | |     hi(j, i) = ht
|   | |     ho(j, i) = 0
|   | +--ELSE
|   | |     hi(j, i) = 0
|   | |     ho(j, i) = ht
|   | +--END IF
|   +--NEXT
+--NEXT


'**************************************************************
' calculate total heat into and out of each space for one complete cycle
'**************************************************************
    hit(1) = 0
    hit(2) = 0
    hit(3) = 0
    hot(1) = 0
    hot(2) = 0
    hot(3) = 0
    hn(1) = 0
    hn(2) = 0
    hn(3) = 0

    ' step through one complete cycle, with i = 1 to 360 degrees
+--FOR i = 1 TO 360
|   ' step through each space, hot, warm, cold
|   +--FOR j = 1 TO 3
|   | ' total heat into each space
|   |     hit(j) = hit(j) + hi(j, i)
|   | ' total heat out of each space
|   |     hot(j) = hot(j) + ho(j, i)
|   | ' total net heat (into and out of each space)
|   |     hn(j) = hn(j) + hi(j, i) + ho(j, i)
|   +--NEXT
```

Ericsson-Ericsson Heat Pump Program

```
+--NEXT

    ' report and store results in file
    LOCATE , 1
    fmt$ = "heat 1=######.## J  2=######.## J  3=######.## J     ### deg"
    ' cycle through 90 degree steps
+--FOR i = 90 TO 360 STEP 90
|       PRINT USING fmt$; hi(1, i) + ho(1, i); hi(2, i) + ho(2, i);
|           hi(3, i) + ho(3, i); i
|       PRINT #1, USING fmt$; hi(1, i) + ho(1, i); hi(2, i) + ho(2, i);
|           hi(3, i) + ho(3, i); i
+--NEXT

    '***********************************************************************
    ' calculated results of heat transferred
    '***********************************************************************
    fmt$ = "heat in  1=######.## J  2=######.## J  3=######.## J"
    PRINT USING fmt$; hit(1); hit(2); hit(3)
    PRINT #1, USING fmt$; hit(1); hit(2); hit(3)
    fmt$ = "heat out 1=######.## J  2=######.## J  3=######.## J"
    PRINT USING fmt$; hot(1); hot(2); hot(3)
    PRINT #1, USING fmt$; hot(1); hot(2); hot(3)
    fmt$ = "net heat 1=######.## J  2=######.## J  3=######.## J"
    PRINT USING fmt$; hn(1); hn(2); hn(3)
    PRINT #1, USING fmt$; hn(1); hn(2); hn(3)

    PRINT USING "net: ######.## J"; hn(1) + hn(2) + hn(3)
    PRINT #1, USING "net: ######.## J"; hn(1) + hn(2) + hn(3)

    '***********************************************************************
    ' calculate mass flow between spaces
    '***********************************************************************
    n1i = 0
    n1o = 0
    n3i = 0
    n3o = 0

    ' step through one complete cycle, with i = 1 to 360 degrees
+--FOR i = 1 TO 360
|   ' determine how much working fluid moved between previous and current
|   ' steps for space 1 (hot)
|       n1 = n(1, i) - n(1, i - 1)
|   ' determine direction of flow
|   +--IF n1 >= 0 THEN
|   |       n1i = n1i + n1
|   +--ELSE
|   |       n1o = n1o - n1
|   +--END IF
|
|   ' determine how much working fluid moved between previous and current
|   ' steps for space 3 (cold)
|       n3 = n(3, i) - n(3, i - 1)
|   ' determine direction of flow
|   +--IF n3 >= 0 THEN
```

220   Appendix E

```
|  |       n3i = n3i + n3
|  +--ELSE
|  |       n3o = n3o - n3
|  +--END IF
+--NEXT

   ' mass flow average (in and out) for hot space (to reduce numerical errors)
   n1 = (n1i + n1o) / 2
   ' mass flow average (in and out) for cold space (to reduce numerical errors)
   n3 = (n3i + n3o) / 2
   ' calculate amount of heat exchanged as gas goes from hot to warm
   hs1 = n1 * 4 * 5.19 * (1000 - 150) * 5 / 9
   ' calculate amount of heat exchanged as gas goes from warm to cold
   hs3 = n3 * 4 * 5.19 * (150 - 32) * 5 / 9

   ' print results
   PRINT USING "mass transfer in/out (1-2): #####.####"; n1;
   PRINT USING "              (2-3): #####.####"; n3
   PRINT #1, USING "mass transfer in/out (1-2): #####.####"; n1;
   PRINT #1, USING "              (2-3): #####.####"; n3

   PRINT USING "regenerator storage (1-2): ######.# J"; hs1;
   PRINT USING "              (2-3): ######.# J"; hs3
   PRINT #1, USING "regenerator storage (1-2): ######.# J"; hs1;
   PRINT #1, USING "              (2-3): ######.# J"; hs3

   '*****************************************************************
   ' calculate work in/out
   '*****************************************************************
   wi = 0
   wo = 0
   ' step through one complete cycle, with i = 1 to 360 degrees
+--FOR i = 1 TO 360
|  ' calculate work
|     w = (hi(1, i) + hi(2, i) + hi(3, i) + ho(1, i) + ho(2, i) + ho(3, i))
|  ' determine direction of flow
|  ' work out of the concept is a positive value;
|  ' work into the concept is a negative value
|  +--IF w < 0 THEN
|  |      wi = wi + w
|  +--ELSE
|  |      wo = wo + w
|  +--END IF
+--NEXT

   ' work average (in and out) (to reduce numerical errors)
   w = (-wi + wo) / 2

   ' print results
   PRINT USING "work storage: ######.# J"; w
   PRINT #1, USING "work storage: ######.# J"; w


   '*****************************************************************
```

Ericsson-Ericsson Heat Pump Program

```
' print out details for file
'••••••••••••••••••••••••••••••••••••••••••••••••••••••••••••••••
  PRINT #1, "volumes throughout cycle"
+--FOR i = 1 TO 360
|    PRINT #1, USING "###    #####.## cm^3"; i; vol(1, i);
|    PRINT #1, USING "    #####.## cm^3"; vol(2, i);
|    PRINT #1, USING "    #####.## cm^3"; vol(3, i)
+--NEXT

  PRINT #1, "pressure throughout cycle"
+--FOR i = 1 TO 360
|    PRINT #1, USING "###   ###.## MPa"; i; pres(i)
+--NEXT

  PRINT #1, "heat transfer in/out of spaces"
+--FOR i = 1 TO 360
|    PRINT #1, USING "###    ####.#### J"; i; hi(1, i) + ho(1, i);
|    PRINT #1, USING "    ####.#### J"; hi(2, i) + ho(2, i);
|    PRINT #1, USING "    ####.#### J"; hi(3, i) + ho(3, i)
+--NEXT

  PRINT #1, "mass transfer in/out of spaces"
  ' step through one complete cycle, with i = 1 to 360 degrees
+--FOR i = 1 TO 360
|    PRINT #1, USING "###"; i;
|    ' step through each space
|  +--FOR j = 1 TO 3
|  |    PRINT #1, USING "   ##.#####"; n(j, i) - n(j, i - 1);
|  +--NEXT
|    PRINT #1,
+--NEXT

  PRINT #1, "work transfer in/out during cycle"
  ' step through one complete cycle, with i = 1 to 360 degrees
+--FOR i = 1 TO 360
|    w = -(hi(1, i) + hi(2, i) + hi(3, i) + ho(1, i) + ho(2, i) + ho(3, i))
|    PRINT #1, USING "###    ######.#### J"; i; w
+--NEXT

  ' close file
  CLOSE
  END
```

Appendix F

Thermal-Compression Vuilleumier Heat Exchanger Analysis Program

The TCVLM.BAS program, written in QuickBASIC version 4.0 (a trademark of Microsoft Corporation), evaluates the influence of realistic heat exchangers and regenerators on the performance of an ideal, isothermal operation, thermal compression Vuilleumier cycle heat pump.

The program starts by determining the loads on the heat exchangers and regenerators needed for an ideal isothermal-cycle operation with the specified cooling capacity and mean cycle pressure and with no dead space. The heat exchangers and regenerators are then sized to meet these loads, according to their design parameters. The dead-space volume is calculated and added to the total system volume. The sizes of the hot and cold working spaces are adjusted to compensate for the decrease in cooling capacity resulting from the dead-space volume. To maintain the specified mean pressure, the amount of working fluid is adjusted, so that a further adjustment of the sizes of the heat exchangers and regenerators is required. The program repeats these calculations until the specified cooling capacity is reached at the specified mean cycle pressure. The number of iterations required depends on the convergence criteria.

The fluid flow friction losses in the heat exchangers and regenerators are also determined. The auxiliary power requirements for these losses are calculated for use in performance evaluation. Pressure differences between spaces that result from the flow friction losses are not considered in determining the cycle's pressure variation with respect to time; that is, pressure is still considered instantaneously uniform for all working fluid. The assumption is that the change in operating characteristics that would result if each space were allowed to have a different pressure is small compared to the overall performance. However, if the design has significant flow friction losses, their effect on the pressures occurring in each space should be considered.

Before executing the program, the following parameters should be specified within the program to properly characterize the design.

Parameter name	Parameter definition	Units
	Working Fluid	
rg	Specific gas constant	kG · m/kg · K
cpg	Specific heat	kcal/kg · K
mw	Molecular weight	kg/kmol
	Thermodynamic Cycle	
tmeanp	Mean cycle pressure	kG/cm
th	Hot-space temperature	°F
thw	Warm-hot-space temperature	°F
tc	Cold-space temperature	°F
tcw	Warm-cold-space temperature	°F
n	Number of cycles per minute	min^{-1}
opvr	Ratio of cold volume to hot volume	None
	Regenerators	
por	Porosity (ratio of open-space volume to total geometric volume)	None
cpr	Specific heat of the matrix	kcal/kg · K
swr	Density of the matrix (including open space)	kg/m^3
sar	Surface density (ratio of area in contact with working fluid, i.e., wetted area, to mass of regenerator matrix)	m^2/kg
rldrh	Hot-regenerator length-to-diameter ratio	None
rldrc	Cold-regenerator length-to-diameter ratio	None
efhro	Hot-regenerator efficiency	None
efcro	Cold-regenerator efficiency	None
	Heat Exchangers	
thinp	Temperature difference of hot space	K
dthwinp	Temperature difference of warm-hot space	K
dtcinp	Temperature difference of cold space	K
dtcwinp	Temperature difference of warm-cold space	K
dzeta	Ratio of free flow area to total frontal area	None
alfa	Ratio of total heat transfer area (wetted area) to total volume	m^2/m^3
forha	Flow passage hydraulic diameter	m
tfins	Fin thickness	cm
hfins	Fin height	cm
kfins	Material heat conductivity	kcal/m · h · K

Parameter name	Parameter definition	Units
fata	Ratio of fin area to total heat transfer area (wetted area)	None
porosity	Ratio of fluid space volume to total geometric volume	None
reh	Hot heat-exchanger Reynolds number	None
rehw	Warm-hot heat-exchanger Reynolds number	None
rec	Cold heat-exchanger Reynolds number	None
recw	Warm-cold heat-exchanger Reynolds number	None
	Convergence Criteria	
iloia	Maximum acceptable variation between successively calculated values of cooling capacity to be met before switching to smaller iteration steps	None
ilofa	Maximum acceptable variation between successively calculated values of cooling capacity to be met to determine whether numerically stable solution has been achieved	None

During execution, the program will prompt the user for values of the following parameters:

Parameter name	Parameter definition	Units
ticopcc	Target cooling capacity of an ideal cycle	kW
hv	Expected volume of hot space	cm^3

The specific value chosen for hv does not affect the final result, but it does affect the number of iterations needed for convergence.

The program displays and prints out the summary results shown below in a data file with the name TCVLM.DAT.

Cycle Characteristics

- Temperature of each space, °C
- Final volume of each space, including void and total volume, cm^3
- Minimum, maximum, and mean cycle pressure, kG/cm^2
- Pressure ratio
- Ideal-cycle heat input and cooling capacity, kW
- Real-cycle heat input and cooling capacity, kW
- Ideal- and real-cycle cooling COP
- Total auxiliary power requirements resulting from flow friction losses, kW

Regenerator Characteristics

- Working fluid Reynolds numbers
- Dimensions, cm
- Total and void volume, cm^3
- Efficiency
- Required heat-exchange rate between regenerator matrix and working fluid, W
- Heat leak, W
- Gas temperature change resulting from regenerator heat leak, °F
- Flow friction pressure losses, kG/m^2
- Flow friction power requirements, W

Heat-Exchanger Characteristics

- Working fluid Reynolds numbers
- Frontal area, cm^2
- Length, cm
- Total and void volume, cm^3
- Capacity, kW
- Efficiency
- Flow friction pressure losses, kG/m^2
- Flow friction power requirements, W

Conversion Factors

1 W = 1 J/s
1 kW = 3413 Btu/h
1 cm = 0.3937 in
1 cm^2 = 0.1550 in^2
1 cm^3 = 0.0610 in^3
1 MPa = 145 psi
X °F = (X-32)/1.8 °C
X K = (X-273.15) °C

1 R = 5/9 K
1 m = 39.37 in
1 kG/cm^2 = 14.22 psi (lb$_f$/in^2)
1 kg/m^3 = 0.0624 lb$_m$/ft^3
1 m^2/kg = 4.882 ft^2/lb$_m$
1 kcal/kg · K = 1 Btu/lb$_m$ · °F
1 kcal/m · h · K = 0.672 Btu · ft/ft^2 · h · °F
1 kG · m/kg · K = 1.823 lb$_f$ · ft/lb$_m$ · R

228   Appendix F

```
' ****************************************************************
' TCVLM.BAS - simulation of the "real" thermal compression Vuilleumier cycle
' ****************************************************************
' ===================== Input Data =========================================
' ****************************************************************
'
CLS : KEY OFF
LOCATE 10, 15: INPUT "Enter target cooling capacity of ideal cycle [kW]"; ticopcc
LOCATE 12, 15: INPUT "Enter expected volume of hot space [cm^3]"; vh
CLS : LOCATE 11, 18: PRINT "calculations in progress..."
'
' ****************************************************************
' ============ input data - thermodynamic cycle ============================
' ****************************************************************
'
rg = 211.9     ' helium gas constant [kGm/kg-K]
cpg = 1.251    ' specific heat of helium [kcal/kg-K]
mw = 4.0026    ' helium mol weight [kg/kmol]
tmeanp = 51    ' helium mean pressure [kG/cm^2]
th = 1000      ' temperature in hot space [F]
thw = 150      ' temperature in hot-warm space [F]
tc = 32        ' temperature in cold space [F]
tcw = 150      ' temperature in cold-warm space [F]
n = 1000       ' number of cycles per minute
opvr = .769    ' optimized ratio of cold volume to hot volume
'
' ****************************************************************
' ============ input data - regenerators ===================================
' ****************************************************************
'
por = .8       ' regenerator porosity
cpr = .12      ' regenerator material specific heat [kcal/kg-K]
swr = 1538     ' regenerator density [kg/m^3]
sar = 9.5      ' regenerator surface density [m^2/kg]
rldrh = .2     ' length to dia. ratio of hot regenerator
rldrc = .2     ' length to dia. ratio of cold regenerator
efhro = .95    ' hot regenerator efficiency
efcro = .95    ' cold regenerator efficiency
'
' ****************************************************************
' ============ input data - heat exchangers ================================
' ****************************************************************
'
dthinp = 30    ' hot heat exchanger differential temp. [K]
dthwinp = 6    ' hot-warm heat exchanger differential temp. [K]
dtcinp = 6     ' cold heat exchanger differential temp. [K]
dtcwinp = 6    ' cold-warm heat exchanger differential temp. [K]
reh = 5000     ' hot heat exchanger flow Reynolds number
rehw = 8000    ' hot-warm heat exchanger flow Reynolds number
rec = 8000     ' cold heat exchanger flow Reynolds number
recw = 8000    ' cold-warm heat exchanger flow Reynolds number
'
' hot, hot-warm, cold, and cold-warm heat exchanger geometry
' fig. 10-91, Kays & London "Compact Heat Exchangers", 1984
```

```
dzeta = .534  ' free flow area/total frontal area of heat exch.
alfa = 587  ' heat transfer area/toal volume [m^2/m^3]
forha = .003632  ' flow passage hydraulic diameter [m]
tfins = .03302  ' fin thickness [cm]
hfins = .75946  ' fin hights [cm]
kfins = 148.8  ' conductivity [Btu/h-ft-F]
fata = .913  ' fin area /total area ratio
hepor = 1 - .235217  ' porosity factor of evaluated heat exchanger geometry Vv/Vt
'
' ••••••••••••••••••••••••••••••••••••••••••••••••••••••••••••••••••••••••
' ============ input data - convergence criteria ===========================
' ••••••••••••••••••••••••••••••••••••••••••••••••••••••••••••••••••••••••
'
iloia = .1  ' accuracy of initial iteration
ilofa = .005  ' accuracy of final iteration
'
' ••••••••••••••••••••••••••••••••••••••••••••••••••••••••••••••••••••••••
' ==================== units conversion ===================================
' ••••••••••••••••••••••••••••••••••••••••••••••••••••••••••••••••••••••••
'
th = ((th - 32) / 1.8) + 273.15  ' [K]
thw = ((thw - 32) / 1.8) + 273.15  ' [K]
tc = ((tc - 32) / 1.8) + 273.15  ' [K]
tcw = ((tcw - 32) / 1.8) + 273.15  ' [K]
vh = vh / 1000000!  ' [m^3]
tfins = tfins / 2.54  ' [in]
hfins = hfins / 2.54  ' [in]
kfins = kfins * .672  ' [Btu/h-ft-F]
'
' ••••••••••••••••••••••••••••••••••••••••••••••••••••••••••••••••••••••••
' ==================== Parameter initialization ===========================
' ••••••••••••••••••••••••••••••••••••••••••••••••••••••••••••••••••••••••
'
pi = 3.141593
' -------------------------------------------------------------------------
tvoidv = 1  ' void volume initial value [cm^3]
tempvd = 150  ' void volume average temperature [K]
' -------------------------------------------------------------------------
rcopx = 1  ' real cooling COP correction factor
iccx = 1  ' ideal cooling capacity correction factor
' -------------------------------------------------------------------------
lla = 1 - iloia  ' low limit of initial iteration accuracy
hla = 1 + iloia  ' high limit of initial iteration accuracy
llafinal = 1 - ilofa  ' low limit of final iteration accuracy
hlafinal = 1 + ilofa  ' high limit of final iteration accuracy
' -------------------------------------------------------------------------
mg = (10000 * tmeanp * 3 * (vh + (vh * opvr))) / (rg * .3 * (th + tc))
  ' initial mass of gas [kg]
'
' ••••••••••••••••••••••••••••••••••••••••••••••••••••••••••••••••••••••••
' ••••••••••••••••••••••••••••••••••••••••••••••••••••••••••••••••••••••••
' ====== Vuilleumier cycle pressure and mass flow calculation ========
' ••••••••••••••••••••••••••••••••••••••••••••••••••••••••••••••••••••••••
```

230   Appendix F

```
' ••••••••••••••••••••••••••••••••••••••••••••••••••••••••••••••••••
'
mhmax = 0: mcmax = 0: mhmin = 100: mcmin = 100: meanpa = 0  ' parameter initialization
' ---------------------------------------------------------------------
vc = vh * opvr  ' calculate cold space volume [m^3]
' ---------------------------------------------------------------------
' calculate constant elements of instantaneous pressure equation
'
aa = (vh / th) + (vh / thw) + (vc / tc) + (vc / tcw) + (2 * ((tvoidv / 1000000!) / tempvd))
bb = (vh / thw) - (vh / th)
cc = (vc / tcw) - (vc / tc)
' ---------------------------------------------------------------------
'
+--FOR alfaa = 1 TO 360
|    alfar = alfaa * (3.141593 / 180)  ' convert degrees to radians
|    press = (2 * mg * rg) / (aa + (bb * COS(alfar)) + (cc * SIN(alfar)))
|       ' calculate system pressure [kG/m^2]
|    mh = (press * (vvh + (.5 * vh * (1 - COS(alfar))))) / (rg * th)  ' calculate hot volume
|    gas mass [kg]
|    mc = (press * (vvc + (.5 * vc * (1 - COS(alfar))))) / (rg * tc)  ' calculate cold
|    volume gas  mass [kg]
|    IF mh > mhmax THEN mhmax = mh  ' find maximum mass in hot space [kg]
|    IF mc > mcmax THEN mcmax = mc  ' find maximum mass in cold space [kg]
|    IF mh < mhmin THEN mhmin = mh  ' find minimum mass in hot space  [kg]
|    IF mc < mcmin THEN mcmin = mc  ' find minimum mass in cold space [kg]
|    meanpa = meanpa + press  '
+--NEXT alfaa

' ---------------------------------------------------------------------
' hot and cold space mass changes  [kmol/cycle]
'
mmh = (mhmax - mhmin) / mw
mmc = (mcmax - mcmin) / mw
' ---------------------------------------------------------------------
'
' hot and cold space mass flow  [lb/sec]
'
mmht = 2 * mmh * mw * (n / 60) * 2.2
mmct = 2 * mmc * mw * (n / 60) * 2.2
' ---------------------------------------------------------------------
'
' maximum and minimum cycle pressure [kG/m^2]
'
pmax = (2 * mg * rg) / (aa - (bb ^ 2 + cc ^ 2) ^ .5)
pmin = (2 * mg * rg) / (aa + (bb ^ 2 + cc ^ 2) ^ .5)
' ---------------------------------------------------------------------
pratio = pmax / pmin  ' calculate cycle pressure ratio
' ---------------------------------------------------------------------
meanp = (meanpa / 360) / 10000  ' calculate mean cycle pressure [kG/cm^2]

' ••••••••••••••••••••••••••••••••••••••••••••••••••••••••••••••••••
' ============== cycle hot and cold space work ===========================
' ••••••••••••••••••••••••••••••••••••••••••••••••••••••••••••••••••
'
```

Thermal-Compression Vuilleumier Heat Exchanger   231

```
' hot and cold space work  [kGm/cycle]
'
qh = -1 * ((2 * 3.141593 * cc * mg * rg * vh) / (bb ^ 2 + cc ^ 2)) * ((1 / ((1 - ((bb ^ 2
+ cc ^ 2) / aa ^ 2)) ^ .5) - 1))
qc = ((2 * 3.141593 * bb * mg * rg * vc) / (bb ^ 2 + cc ^ 2)) * ((1 / ((1 - ((bb ^ 2 + cc
^ 2) / aa ^ 2)) ^ .5) - 1))
' -----------------------------------------------------------------------
qc = (qc + qcp) / 2: qcp = qc
qh = (qh + qhp) / 2: qhp = qh
'
'**********************************************************************
'======================= regenerators ================================
'**********************************************************************
'
' calculate inlet and outlet gas density for hot and cold regenerator [kg/m^3]
'
roorhin = (meanp * 10000) / (rg * (thw + deltath))
roorhout = (meanp * 10000) / (rg * th)
roorcin = (meanp * 10000) / (rg * tc)
roorcout = (meanp * 10000) / (rg * (tcw - deltatc))
' -----------------------------------------------------------------------
' calculate regenerator size needed to match requested efficiency
' -----------------------------------------------------------------------
steph = 8: steph0 = 10: stepc = 10: stepc0 = 10 ' initial step of size adjustment
' -----------------------------------------------------------------------
' diameter of hot and cold regenerator  [cm]
'
1860    dhr = dhr + steph
        dcr = dcr + stepc
' -----------------------------------------------------------------------
' length of hot and cold regenerator  [cm]
'
lhr = rldrh * dhr
lcr = rldrc * dcr
' -----------------------------------------------------------------------
' frontal area of hot and cold regenerator  [cm^2]
'
afh = (pi * dhr ^ 2) / 4
afc = (pi * dcr ^ 2) / 4
' -----------------------------------------------------------------------
' volume of hot and cold regenerator  [m^3]
'
vrh = (lrh * afh) / 1000000!
vrc = (lrc * afc) / 1000000!
' -----------------------------------------------------------------------
avr = sar * swr ' surface density on a volume base [m^2/m^3] of regen.
' -----------------------------------------------------------------------
'
' mass flow through hot and cold regenerator [kg/sec-m^2]
```

232 Appendix F

```
grh = ((mmht / 2.2) / (afh * por)) * 10000
grc = ((mmct / 2.2) / (afc * por)) * 10000
' -----------------------------------------------------------------------
' gas average dynamic viscosity in hot and cold regenerator  [lb/ft-sec]
'
viscrh = 10.5 * ((((((((th + thw) / 2) - 273.15) * 1.8) + 32 + 459.69) ^ .617) / 10000) /
  3600
viscrc = 10.5 * ((((((((tc + tcw) / 2) - 273.15) * 1.8) + 32 + 459.69) ^ .617) / 10000) /
  3600
' -----------------------------------------------------------------------
'
' average coefficient of heat transfer in hot and cold regenerator [Btu/sec-ft^2-R]
'
hrh = .421 * cpg * (viscrh * avr * .3048) ^ .45 * (mmht / (afh / 929)) ^ .55
hrc = .421 * cpg * (viscrc * avr * .3048) ^ .45 * (mmct / (afc / 929)) ^ .55
' -----------------------------------------------------------------------
' inefficiency of hot and cold regenerator
'
eh = (2 * mmht * cpg) / (hrh * avr * (10.76 / 35.3) * (afh / 929) * (lhr / 30.48))
ec = (2 * mmct * cpg) / (hrc * avr * (10.76 / 35.3) * (afc / 929) * (lcr / 30.48))
' -----------------------------------------------------------------------
'
' step of hot and cold regenerator diameter sizing [cm]
'
steph = steph0 * (eh - (1 - efhro))
stepc = stepc0 * (ec - (1 - efcro))
' -----------------------------------------------------------------------
IF eh < (1 - efhro) - .001 OR eh > (1 - efhro) + .001 THEN GOTO 1860
IF ec < (1 - efcro) - .001 OR ec > (1 - efcro) + .001 THEN GOTO 1860
' -----------------------------------------------------------------------
' ======================================================================
'
' energy storage requirement for hot and cold regenerator [J/cycle]
'
tqrh = mmh * mw * cpg * (th - thw) * 4190
tqrc = mmc * mw * cpg * (tcw - tc) * 4190
' -----------------------------------------------------------------------
'
' energy leak in hot and cold regenerator [J/cycle]
'
leakh = eh * tqrh
leakc = ec * tqrc
' -----------------------------------------------------------------------
'
' gas temperature change due to hot and cold regenerator inefficiency [K]
'
deltath = leakh / (mmh * mw * cpg * 4190)
deltatc = leakc / (mmc * mw * cpg * 4190)
' -----------------------------------------------------------------------
' hot and cold regenerator gas flow Reynolds number
```

```
rehr = (4 * mmht) / (avr * (10.76 / 35.3) * (afh / 929) * viscrh)
recr = (4 * mmct) / (avr * (10.76 / 35.3) * (afc / 929) * viscrc)
' ---------------------------------------------------------------------
'
' hot and cold regenerator gas flow friction factor
' fig. 10-100, Kays & London "Compact Heat Exchangers", 1984
'
fprimrh = .5846342 - 3.050584E-02 * LOG(rehr)
fprimrc = .5846342 - 3.050584E-02 * LOG(recr)
' ---------------------------------------------------------------------
'
' hot and cold regenerator flow friction pressure drop [kG/m^2]
'
dph = ((grh ^ 2 / (2 * 9.807)) * (1 / roorhin)) * ((1 + por ^ 2) * ((roorhin / roorhout)
 - 1) + (fprimrh * ((avr * vrh) / (por * (afh / 10000))) * (roorhin / ((roorhin +
 roorhout) / 2))))
dpc = ((grc ^ 2 / (2 * 9.807)) * (1 / roorcin)) * ((1 + por ^ 2) * ((roorcin / roorcout)
 - 1) + (fprimrc * ((avr * vrc) / (por * (afc / 10000))) * (roorcin / ((roorcin +
 roorcout) / 2))))
' ---------------------------------------------------------------------
'
' hot and cold regenerator flow friction power requirements [hp]
'
hph = (dph * .205 * mmht) / (550 * (((roorhin + roorhout) / 2) / 16.02))
hpc = (dpc * .205 * mmct) / (550 * (((roorcin + roorcout) / 2) / 16.02))
'
'*****************************************************************
' =================== heat exchangers ============================
'*****************************************************************
'
' hot, hot-warm, cold, and cold-warm heat exchanger load [kcal]
'
totqh  = ((qh * n * 60) / 427) + ((leakh * (n / 60)) * .86)
totqhw = ((qh * n * 60) / 427) + ((leakh * (n / 60)) * .86)
totqc  = ((qc * n * 60) / 427) - ((leakc * (n / 60)) * .86)
totqcw = ((qc * n * 60) / 427) - ((leakc * (n / 60)) * .86)
' ---------------------------------------------------------------------
'
' hot, hot-warm, cold, and cold-warm heat exchanger diff. temperature [K]
'
dth  = dthinp + deltath
dthw = dthwinp + deltath
dtc  = dtcinp - deltatc
dtcw = dtcwinp - deltatc
' ---------------------------------------------------------------------
'
' average dynamic viscosity of helium flowing thrugh hot, hot-warm, cold, and cold-warm
 heat exchanger [lb/ft-sec]
'
visrh  = 10.5 * ((((((th - deltath) - 273.15) * 1.8) + 32 + 459.69) ^ .617) / 10000) / 3600
visrhw = 10.5 * ((((((thw + deltath) - 273.15) * 1.8) + 32 + 459.69) ^ .617) / 10000) /
 3600
visrc  = 10.5 * ((((((tc + deltatc) - 273.15) * 1.8) + 32 + 459.69) ^ .617) / 10000) / 3600
```

234   Appendix F

```
visrcw = 10.5 * ((((((tcw - deltatc) - 273.15) * 1.8) + 32 + 459.69) ^ .617) / 10000) /
3600
```
' ---

' average thermal conductivity of helium flowing through hot, hot-warm, cold, and
' cold-warm heat exchanger [kcal/m-h-K]

```
lambgh   = 1.488 * (8.231081E-02 + 8.770272E-05 * ((((th - deltath) - 273.15) * 1.8) + 32))
lambghw  = 1.488 * (8.231081E-02 + 8.770272E-05 * ((((thw + deltath) - 273.15) * 1.8) + 32))
lambgc   = 1.488 * (8.231081E-02 + 8.770272E-05 * ((((tc + deltatc) - 273.15) * 1.8) + 32))
lambgcw  = 1.488 * (8.231081E-02 + 8.770272E-05 * ((((tcw - deltatc) - 273.15) * 1.8) + 32))
```
' ---

' average Prandtl number of helium flowing through hot, hot-warm, cold, and cold-warm
' heat exchanger

```
prnh  = (cpg * visrh  * 1.488) / (lambgh  / 3600)
prnhw = (cpg * visrhw * 1.488) / (lambghw / 3600)
prnc  = (cpg * visrc  * 1.488) / (lambgc  / 3600)
prncw = (cpg * visrcw * 1.488) / (lambgcw / 3600)
```
' ---

' required frontal area of hot, hot-warm, cold, and cold-warm heat exchanger [m^2]

```
efh  = ((mmht / 2.2) * forha) / (reh  * visrh  * 1.488 * dzeta)
efhw = ((mmht / 2.2) * forha) / (rehw * visrhw * 1.488 * dzeta)
efc  = ((mmct / 2.2) * forha) / (rec  * visrc  * 1.488 * dzeta)
efcw = ((mmct / 2.2) * forha) / (recw * visrcw * 1.488 * dzeta)
```
' ---

' average mass flow through hot, hot-warm, cold, and cold-warm heat exchanger [kg/sec-m^2]

```
gh  = (mmht / 2.2) / (efh  * dzeta)
ghw = (mmht / 2.2) / (efhw * dzeta)
gc  = (mmct / 2.2) / (efc  * dzeta)
gcw = (mmct / 2.2) / (efcw * dzeta)
```
' ---

' hot, hot-warm, cold, and cold-warm heat exchanger gas flow StPr^2/3
' factor, fig. 10-91, Kays & London "Compact Heat Exchangers", 1984

```
heqh  = .1644257 * reh  ^ -.3980834
heqhw = .1644257 * rehw ^ -.3980834
heqc  = .1644257 * rec  ^ -.3980834
heqcw = .1644257 * recw ^ -.3980834
```
' ---

' hot, hot-warm, cold, and cold-warm heat exchanger gas flow friction
' factor, fig. 10-91, Kays & London "Compact Heat Exchangers", 1984

```
fprimh  = .1300191 * reh  ^ -.2127139
fprimhw = .1300191 * rehw ^ -.2127139
fprimc  = .1300191 * rec  ^ -.2127139
```

fprimcw = .1300191 * recw ^ -.2127139
'--
'
' coefficient of heat transfer of hot, hot-warm, cold, and cold-warm heat exchanger
' (helium side), [kcal/h-m^2-K]
'
hh = (heqh / prnh ^ (2 / 3)) * gh * cpg * 3600
hhw = (heqhw / prnhw ^ (2 / 3)) * ghw * cpg * 3600
hc = (heqc / prnc ^ (2 / 3)) * gc * cpg * 3600
hcw = (heqcw / prncw ^ (2 / 3)) * gcw * cpg * 3600
'--
'
' hot, hot-warm, cold, and cold-warm heat exchanger geometric parameter
' needed in evaluation of fin heat transfer effectiveness factor
' (helium side), fig. 2-13, Kays & London "Compact Heat Exchangers", 1984
'
mcorh = SQR((2 * (hh / 4.88)) / (kfins * (tfins / 12))) * (hfins / 12)
mcorhw = SQR((2 * (hhw / 4.88)) / (kfins * (tfins / 12))) * (hfins / 12)
mcorc = SQR((2 * (hc / 4.88)) / (kfins * (tfins / 12))) * (hfins / 12)
mcorcw = SQR((2 * (hcw / 4.88)) / (kfins * (tfins / 12))) * (hfins / 12)
'--
'
' hot, hot-warm, cold, and cold-warm heat exchanger fin heat transfer
' effectiveness factor (helium side), fig. 2-13, Kays & London
' "Compact Heat Exchangers", 1984
'
fineffh = 1.043665 - .3191194 * mcorh - 9.974898E-02 * mcorh ^ 2 + 4.403333E-02 * mcorh ^ 3
fineffhw = 1.043665 - .3191194 * mcorhw - 9.974898E-02 * mcorhw ^ 2 + 4.403333E-02 *
 mcorhw ^ 3
fineffc = 1.043665 - .3191194 * mcorc - 9.974898E-02 * mcorc ^ 2 + 4.403333E-02 * mcorc ^ 3
fineffcw = 1.043665 - .3191194 * mcorcw - 9.974898E-02 * mcorcw ^ 2 + 4.403333E-02 *
 mcorcw ^ 3
'--
'
' overall surface effectiveness of hot, hot-warm, cold, and cold-warm heat exchanger
' (helium side)
'
ofeffh = 1 - (fata * (1 - fineffh))
ofeffhw = 1 - (fata * (1 - fineffhw))
ofeffc = 1 - (fata * (1 - fineffc))
ofeffcw = 1 - (fata * (1 - fineffcw))
'--
'
' length of hot, hot-warm, cold, and cold-warm heat exchanger [m]
'
elh = totqh / (hh * ofeffh * dth * efh * alfa)
elhw = totqhw / (hhw * ofeffhw * dthw * efhw * alfa)
elc = totqc / (hc * ofeffc * dtc * efc * alfa)
elcw = totqcw / (hcw * ofeffcw * dtcw * efcw * alfa)
'--
'
' total volume of hot, hot-warm, cold, and cold-warm heat exchanger [m^3]
'
evh = elh * efh

```
evhw = elhw * efhw
evc  = elc * efc
evcw = elcw * efcw
' --------------------------------------------------------------------------
'
' void volume of hot, hot-warm, cold, and cold-warm heat exchanger [m^3]
'
vvh  = hepor * evh
vvhw = hepor * evhw
vvc  = hepor * evc
vvcw = hepor * evcw
' --------------------------------------------------------------------------
'
' average inlet gas density in hot, hot-warm, cold, and cold-warm heat exchanger [kg/m^3]
'
roohin  = (meanp * 10000) / (rg * (th - deltath))
roohwin = (meanp * 10000) / (rg * (thw + deltath))
roocin  = (meanp * 10000) / (rg * (tc + deltatc))
roocwin = (meanp * 10000) / (rg * (tcw + deltatc))
' --------------------------------------------------------------------------
'
' average outlet gas density in hot, hot-warm, cold, and cold-warm heat exchanger [kg/m^3]
'
roohout  = (meanp * 10000) / (rg * th)
roohwout = (meanp * 10000) / (rg * thw)
roocout  = (meanp * 10000) / (rg * tc)
roocwout = (meanp * 10000) / (rg * tcw)
' --------------------------------------------------------------------------
'
' hot, hot-warm, cold, and cold-warm heat exchanger flow friction pressure drops [kG/m^2]
'
hedph = ((gh ^ 2 / (2 * 9.807)) * (1 / roohin)) * ((1 + dzeta ^ 2) * ((roohin / roohout)
   - 1) + (fprimh * ((alfa * evh) / (dzeta * efh)) * (roohin / ((roohin + roohout) / 2))))
hedphw = ((ghw ^ 2 / (2 * 9.807)) * (1 / roohwin)) * ((1 + dzeta ^ 2) * ((roohwin /
   roohwout) - 1) + (fprimhw * ((alfa * evhw) / (dzeta * efhw)) * (roohwin / ((roohwin +
   roohwout) / 2))))
hedpc = ((gc ^ 2 / (2 * 9.807)) * (1 / roocin)) * ((1 + dzeta ^ 2) * ((roocin / roocout)
   - 1) + (fprimc * ((alfa * evc) / (dzeta * efc)) * (roocin / ((roocin + roocout) / 2))))
hedpcw = ((gcw ^ 2 / (2 * 9.807)) * (1 / roocwin)) * ((1 + dzeta ^ 2) * ((roocwin /
   roocwout) - 1) + (fprimcw * ((alfa * evcw) / (dzeta * efcw)) * (roocwin / ((roocwin +
   roocwout) / 2))))
' --------------------------------------------------------------------------
'
' hot, hot-warm, cold, cold-warm heat exchangers flow friction power losses [hp]
'
hehph  = (hedph * .205 * mmht) / (550 * (((roohin + roohout) / 2) / 16.02))
hehphw = (hedphw * .205 * mmht) / (550 * (((roohwin + roohwout) / 2) / 16.02))
hehpc  = (hedpc * .205 * mmct) / (550 * (((roocin + roocout) / 2) / 16.02))
hehpcw = (hedpcw * .205 * mmct) / (550 * (((roocwin + roocwout) / 2) / 16.02))
'
' ****************************************************************
' ================== summary =========================================
' ****************************************************************
'
' total flow friction energy losses [j/cycle]
```

Thermal-Compression Vuilleumier Heat Exchanger

```
        tdeltap = ((hph + hpc + hehph + hehphw + hehpc + hehpcw) * 744.444) / (n / 60)
        '
        ' total void volume  [cm^3]
        tvoidv = (afh * lhr * por) + (afc * lcr * por) + ((vvh + vvhw + vvc + vvcw) * 1000000!)
        '
        ' average temperature of void volume [K]
        tempvd = (tvoidv / 1000000!) / ((((afh * lhr * por) / 1000000!) / ((th + thw) / 2)) +
           (((afc * lcr * por) / 1000000!) / ((tc + tcw) / 2)) + (vvh / th) + (vvhw / thw) + (vvc /
           tc) + (vvcw / tcw))
        '
        ' ideal cycle cooling COP
        icop = qc / qh
        '
        ' ideal cycle heat input [kW]
        icophin = ((qh * 9.807) * (n / 60)) / 1000
        '
        ' ideal cycle cooling capacity [kW]
        icopcc = ((qc * 9.807) * (n / 60)) / 1000
        '
        ' real cycle cooling COP
        rcop = totqc / totqh
        '
        ' real cycle heat input  [kW]
        rhin = totqh / 860
        '
        ' real cycle cooling capacity  [kW]
        rcc = totqc / 860
        '
        ' ************************************************************
        ' ==== adjust gas quantity, hot volume to meet convergence criteria =======
        ' ************************************************************
        '
        ' check if real cycle cooling COP within iteration accuracy limits
        '
        xrcop = rcop / rcopx: rcopx = rcop
        IF (xrcop >= lla) AND (xrcop <= hla) THEN rcopflag = 1 ELSE rcopflag = 0
        IF rcopflag = 0 THEN GOTO 5030
        '
        ' check if real cycle mean pressure within limits, adjust mass of gas if needed
        '
        xmeanp = meanp / tmeanp: mg = mg / xmeanp
        IF (xmeanp >= lla) AND (xmeanp <= hla) THEN meanpflg = 1 ELSE meanpflg = 0
        IF meanpflg = 0 THEN GOTO 5030
        '
        ' check if ideal cycle cooling capacity within limits, adjust hot space volume if needed
        '
        xicopcc = icopcc / ticopcc: vh = vh / xicopcc
        IF (xicopcc >= lla) AND (xicopcc <= hla) THEN iccflag = 1 ELSE iccflag = 0
        '
        IF (aflag = 1) THEN GOTO 5020
        IF (iccflag = 1) THEN lla = llafinal: hla = hlafinal: aflag = 1: iccflag = 0
        IF (meanpflg = 1) AND (iccflag = 1) THEN GOTO 5090
5020    ' ----------------------------------------------------------------
5030    LOCATE 17, 40: PRINT SPACE$(39)
```

238 Appendix F

```
            LOCATE 17, 40: PRINT icopcc; " ideal cycle cooling capacity "
            LOCATE 19, 40: PRINT SPACE$(39)
            LOCATE 19, 40: PRINT rcop; " real cycle COP"
            LOCATE 21, 40: PRINT SPACE$(39)
            LOCATE 21, 40: PRINT mg * 1000; " gas mass"
            LOCATE 23, 40: PRINT SPACE$(39)
            LOCATE 23, 40: PRINT vh * 1000000!; " hot space volume"
            GOTO 1040
 5090       CLS
            '
            ' ••••••••••••••••••••••••••••••••••••••••••••••••••••••••••••••••••
            ' =============== print final results =================================
            ' ••••••••••••••••••••••••••••••••••••••••••••••••••••••••••••••••••
            '
            OPEN "tcvlm.dat" FOR OUTPUT AS 1
            PRINT #1, "•••••••••••••••••••••••••••••••••••••••••••••••••••": PRINT #1,
            PRINT "       ------------- THERMAL COMPRESSION VUILLEUMIER ---------------"
            PRINT #1, "      ------------- THERMAL COMPRESSION VUILLEUMIER ---------------"
            PRINT #1, "•••••••••••••••••••••••••••••••••••••••••••••••••••": PRINT #1,
            PRINT "-------------------- cycle parameters --------------------------"
            PRINT #1, "-------------------- cycle parameters --------------------------"
            PRINT "Th=";  : PRINT USING "####.#"; th - 273.15;
            PRINT #1, "Th=";  : PRINT #1, USING "####.#"; th - 273.15;
            PRINT "  Thw=";  : PRINT USING "####.#"; thw - 273.15;
            PRINT #1, "  Thw=";  : PRINT #1, USING "####.#"; thw - 273.15;
            PRINT "  Tc=";  : PRINT USING "####.#"; tc - 273.15;
            PRINT #1, "  Tc=";  : PRINT #1, USING "####.#"; tc - 273.15;
            PRINT "  Tcw=";  : PRINT USING "####.#"; tcw - 273.15; : PRINT "  [C]"
            PRINT #1, "  Tcw=";  : PRINT #1, USING "####.#"; tcw - 273.15; : PRINT #1, "  [C]"
            PRINT "Vh=";  : PRINT USING "######.#"; vh * 1000000!;
            PRINT #1, "Vh=";  : PRINT #1, USING "######.#"; vh * 1000000!;
            PRINT "  Vc=";  : PRINT USING "######.#"; vc * 1000000!;
            PRINT #1, "  Vc=";  : PRINT #1, USING "######.#"; vc * 1000000!;
            PRINT "  Vt=";  : PRINT USING "######.#"; tvoidv + ((vh + vc) * 1000000!); :
              PRINT "  [cm^3]"
            PRINT #1, "  Vt=";  : PRINT #1, USING "######.#"; tvoidv + ((vh + vc) * 1000000!); :
              PRINT #1, "  [cm^3]"
            PRINT "void volume=";  : PRINT USING "######.#"; tvoidv; : PRINT "  [cm^3]";
            PRINT #1, "void volume=";  : PRINT #1, USING "######.#"; tvoidv; : PRINT #1, "  [cm^3]";
            PRINT "    void volume ratio=";  : PRINT USING "#.###"; tvoidv / (tvoidv + ((vh + vc) *
              1000000!))
            PRINT #1, "    void volume ratio=";  : PRINT #1, USING "#.###"; tvoidv / (tvoidv + ((vh +
              vc) * 1000000!))
            PRINT "Pmax=";  : PRINT USING "####.##"; pmax / 10000;
            PRINT #1, "Pmax=";  : PRINT #1, USING "####.##"; pmax / 10000;
            PRINT "  Pmin=";  : PRINT USING "####.##"; pmin / 10000;
            PRINT #1, "  Pmin=";  : PRINT #1, USING "####.##"; pmin / 10000;
            PRINT "  Pmean=";  : PRINT USING "####.##"; meanp; : PRINT "  [kG/cm^2]"
            PRINT #1, "  Pmean=";  : PRINT #1, USING "####.##"; meanp; : PRINT #1, "  [kG/cm^2]"
            PRINT "Pr=";  : PRINT USING "#.##"; pratio;
            PRINT #1, "Pr=";  : PRINT #1, USING "#.##"; pratio;
            PRINT "  Mg=";  : PRINT USING "####.#"; mg * 1000; : PRINT "  [g]";
            PRINT #1, "  Mg=";  : PRINT #1, USING "####.#"; mg * 1000; : PRINT #1, "  [g]";
            PRINT "  N=";  : PRINT USING "####."; n; : PRINT "  [1/min]"
```

Thermal-Compression Vuilleumier Heat Exchanger

```
PRINT #1, " N=";  : PRINT #1, USING "####.";  n;  : PRINT #1, " [1/min]"
PRINT "----------------------------------- ideal cycle --- real cycle ------"
PRINT #1, "----------------------------------- ideal cycle --- real cycle ------"
PRINT " cooling COP                   ";  : PRINT USING "###########.###";  icop; rcop
PRINT #1, " cooling COP               ";  : PRINT #1, USING "###########.###";  icop;
    rcop
PRINT " heat input          [kW]      ";  : PRINT USING "###########.###";  icophin; rhin
PRINT #1, " heat input      [kW]      ";  : PRINT #1, USING "###########.###";  icophin;
    rhin
PRINT " cooling capacity    [kW]      ";  : PRINT USING "###########.###";  icopcc; rcc
PRINT #1, " cooling capacity  [kW]    ";  : PRINT #1, USING "###########.###";  icopcc;
    rcc
PRINT " flow fric. losses   [kW]      ";  : PRINT USING "###########.###";  xxxxx;
    (tdeltap * (n / 60)) / 1000
PRINT #1, " flow fric. losses  [kW]   ";  : PRINT #1, USING "###########.###";  xxxxx;
    (tdeltap * (n / 60)) / 1000
PRINT "---------------------- regenerators ----------------------------------"
PRINT #1, "---------------------- regenerators ----------------------------------"
PRINT "----------------------------------- hot ----------- cold-------------"
PRINT #1, "----------------------------------- hot ----------- cold-------------"
PRINT " Reynold number                ";  : PRINT USING "###########.##";  rehr; recr
PRINT #1, " Reynold number            ";  : PRINT #1, USING "###########.##";  rehr; recr
PRINT " diameter            [cm]      ";  : PRINT USING "###########.##";  dhr; dcr
PRINT #1, " diameter         [cm]     ";  : PRINT #1, USING "###########.##";  dhr; dcr
PRINT " length              [cm]      ";  : PRINT USING "###########.##";  lhr; lcr
PRINT #1, " length           [cm]     ";  : PRINT #1, USING "###########.##";  lhr; lcr
PRINT " volume              [cm^3]    ";  : PRINT USING "###########.##";  afh * lhr;
    afc * lcr
PRINT #1, " volume           [cm^3]   ";  : PRINT #1, USING "###########.##";
    afh * lhr; afc * lcr
PRINT " void volume         [cm^3]    ";  : PRINT USING "###########.##";  afh * lhr * por;
    afc * lcr * por
PRINT #1, " void volume      [cm^3]   ";  : PRINT #1, USING "###########.##";
    afh * lhr * por; afc * lcr * por
PRINT " efficiency                    ";  : PRINT USING "###########.##";  1 - eh; 1 - ec
PRINT #1, " efficiency                ";  : PRINT #1, USING "###########.##";  1 - eh;
    1 - ec
PRINT " heat exchng reg/gas [kW]      ";  : PRINT USING "###########.##";
    (tqrh * (n / 60)) / 1000; (tqrc * (n / 60)) / 1000
PRINT #1, " heat exchng reg/gas [kW]  ";  : PRINT #1, USING "###########.##";
    (tqrh * (n / 60)) / 1000; (tqrc * (n / 60)) / 1000
PRINT " heat leak           [kW]      ";  : PRINT USING "###########.##";
    (leakh * (n / 60)) / 1000; (leakc * (n / 60)) / 1000
PRINT #1, " heat leak        [kW]     ";  : PRINT #1, USING "###########.##";
    (leakh * (n / 60)) / 1000; (leakc * (n / 60)) / 1000
PRINT " gas temp. change    [K]       ";  : PRINT USING "###########.##";  deltath; deltatc
PRINT #1, " gas temp. change [K]      ";  : PRINT #1, USING "###########.##";  deltath;
    deltatc
PRINT " pressure drop       [kG/m^2]  ";  : PRINT USING "###########.##";  dph; dpc
PRINT #1, " pressure drop    [kG/m^2] ";  : PRINT #1, USING "###########.##";  dph; dpc
PRINT " flow fric. power    [W]       ";  : PRINT USING "###########.##";  hph * 744.444;
    hpc * 744.444
PRINT #1, " flow fric. power [W]      ";  : PRINT #1, USING "###########.##";
    hph * 744.444; hpc * 744.444
```

Appendix F

```
PRINT "----------------------- heat exchangers ---------------------------"
PRINT #1, "----------------------- heat exchangers ---------------------------"
PRINT "-------------------------------- hot -- hot-warm -- cold -- cold-warm --"
PRINT #1, "-------------------------------- hot -- hot-warm -- cold -- cold-warm --"
PRINT " Reynolds number           "; : PRINT USING "######.## "; reh; rehw; rec; recw
PRINT #1, " Reynolds number           "; : PRINT #1, USING "######.## "; reh; rehw;
  rec; recw
PRINT " frontal area       [cm^2]  "; : PRINT USING "######.## "; efh * 10000;
  efhw * 10000; efc * 10000; efcw * 10000
PRINT #1, " frontal area       [cm^2]  "; : PRINT #1, USING "######.## "; efh * 10000;
  efhw * 10000; efc * 10000; efcw * 10000
PRINT " length             [cm]    "; : PRINT USING "######.## "; elh * 100;
  elhw * 100; elc * 100; elcw * 100
PRINT #1, " length             [cm]    "; : PRINT #1, USING "######.## "; elh * 100;
  elhw * 100; elc * 100; elcw * 100
PRINT " volume             [cm^3]  "; : PRINT USING "######.## "; evh * 1000000!;
  evhw * 1000000!; evc * 1000000!; evcw * 1000000!
PRINT #1, " volume             [cm^3]  "; : PRINT #1, USING "######.## ";
  evh * 1000000!; evhw * 1000000!; evc * 1000000!; evcw * 1000000!
PRINT " void volume        [cm^3]  "; : PRINT USING "######.## "; vvh * 1000000!;
  vvhw * 1000000!; vvc * 1000000!; vvcw * 1000000!
PRINT #1, " void volume        [cm^3]  "; : PRINT #1, USING "######.## ";
  vvh * 1000000!; vvhw * 1000000!; vvc * 1000000!; vvcw * 1000000!
PRINT " heat capacity      [kW]    "; : PRINT USING "######.## "; (totqh / .86) / 1000;
  (totqhw / .86) / 1000; (totqc / .86) / 1000; (totqcw / .86) / 1000
PRINT #1, " heat capacity      [kW]    "; : PRINT #1, USING "######.## ";
  (totqh / .86) / 1000; (totqhw / .86) / 1000; (totqc / .86) / 1000; (totqcw / .86) / 1000
PRINT " overall efficiency         "; : PRINT USING "######.## "; ofeffh; ofeffhw;
  ofeffc; ofeffcw
PRINT #1, " overall efficiency         "; : PRINT #1, USING "######.## "; ofeffh;
  ofeffhw; ofeffc; ofeffcw
PRINT " pressure drop      [kG/m^2] "; : PRINT USING "######.## "; hedph; hedphw; hedpc;
  hedpcw
PRINT #1, " pressure drop      [kG/m^2] "; : PRINT #1, USING "######.## "; hedph;
  hedphw; hedpc; hedpcw
PRINT " flow fric. power   [W]     "; : PRINT USING "######.## "; hehph * 744.444;
  hehphw * 744.444; hehpc * 744.444; hehpcw * 744.444
PRINT #1, " flow fric. power   [W]     "; : PRINT #1, USING "######.## ";
  hehph * 744.444; hehphw * 744.444; hehpc * 744.444; hehpcw * 744.444
CLOSE
END
```

Appendix G

Glossary

ADIABATIC A type of thermodynamic process in which no heat transfer occurs between the working fluid and the environment.

ALPHA CONFIGURATION A configuration for classifying Stirling machines consisting of two piston-in-cylinder combinations, one for the expansion space and the other for the compression space.

AUXILIARY POWER The power required for mechanical operation of a system, external to the thermodynamic heat and work; usually a result of mechanical losses and the movement of fluids for heat transfer.

BETA CONFIGURATION A configuration for classifying Stirling machines consisting of a single cylinder shared by a displacer and a piston, usually with the expansion space formed between the displacer and the cylinder end, and the compression space formed between the displacer and the piston.

BRAYTON CYCLE A theoretical thermodynamic cycle consisting of four steps as a refrigerator: adiabatic expansion, constant pressure heating, adiabatic compression, and constant pressure cooling. The order is reversed for an engine. It is also referred to as a gas turbine cycle, or as Joule or Ericsson cycles in Europe.

CARNOT CYCLE An ideal thermodynamic cycle consisting of four steps as a refrigerator: isentropic expansion, isothermal expansion, isentropic compression, and isothermal compression. The order of the steps is reversed for an engine.

CLAUSIUS-RANKINE CYCLE The European identification of the Rankine cycle; also referred to as the Clausius cycle.

CLOSED CYCLE A thermodynamic cycle in which all the refrigerant is recirculated.

COEFFICIENT OF PERFORMANCE, COP A figure of merit that describes the performance of refrigerators. It is defined differently for heating and cooling applications (see Heating COP or Cooling COP).

COLD-END TEMPERATURE The temperature at which a refrigerator or the refrigerator segment of a heat-activated heat pump (engine-refrigerator) is absorbing heat; usually the lowest heat pump temperature.

COMFORT HEATING OR COOLING The heating or cooling of an indoor environment for the comfort of the occupants.

COMPRESSION A thermodynamic process resulting in a net increase in pressure of a working fluid; it may or may not involve a reduction of the volume occupied by the working fluid. If done by isothermal volume reduction, the working fluid rejects heat.

COMPRESSION RATIO The ratio of the maximum working fluid volume to the minimum working fluid volume. Note: this is a different definition than that used in compressor technology.

COMPRESSION SPACE The space associated with the compression process, and, in Stirling and similar equipment, with the rejection of heat from the working fluid undergoing compression.

CONCEPT An idea for combining equipment components (e.g., compressor, heat exchanger) into a system, such as a heat pump.

CONSTRAINED PISTON (OR DISPLACER) A piston (or displacer) whose motion is controlled by a mechanical link (or equivalent) to produce the desired operation.

COOLING COP, COP_c The ratio of the heat absorbed at the low-temperature heat exchanger of a refrigerator to the net energy input driving the refrigeration cycle. The energy units in the numerator and the denominator are the same so COPs are dimensionless and without units.

CRANKSHAFT A mechanical linkage to which pistons or displacers are attached, which controls or constrains their motion; usually rotary in operation.

CYCLE The sequence of thermodynamic states of a working fluid as it passes through an engine or refrigerator.

CYLINDER The physical shell in which a piston or displacer resides. It forms one of the surfaces containing the working fluid.

DEAD SPACE See Dead-Space Volume.

DEAD-SPACE VOLUME The volume of the system that contains working fluid but does not go through a volume change to drive the cycle. This is usually located in heat exchangers, regenerators, and connecting channels.

DEAD-SPACE VOLUME FRACTION The ratio of the dead-space volume to the total volume.

DEAD VOLUME See Dead-Space Volume.

DISPLACER A moving element that has an intended temperature difference but no intended pressure difference between opposite sides. A displacer moves working fluid between working spaces, which are usually at different temperatures. A real displacer may have some pressure difference across it due to flow losses in passages connecting the working spaces.

EFFECTIVENESS A measure of efficiency for heat exchangers, usually computed as the ratio of the actual difference in temperature of a fluid due to heat exchange to the ideal or total temperature difference possible.

ENGINE A device that accepts heat at a higher temperature, rejects a smaller amount of heat at a lower temperature, and produces work.

EQUILIBRIUM A condition in which the flows and forces are in balance. In simulations, parameters are iteratively varied in search of this condition.

ERICSSON CYCLE An ideal thermodynamic cycle consisting of four steps as a refrigerator: isobaric expansion, isothermal expansion, isobaric compression, and isothermal compression. The order of the steps is reversed for an engine.

EXPANDER A flow restriction or a mechanical compressor operating in a reverse manner to expand a working fluid rather than to compress it.

EXPANSION A thermodynamic process resulting in a net decrease in pressure of a working fluid; it may or may not involve an increase of the volume occupied by the working fluid. If done by isothermal volume increase, the working fluid absorbs heat.

EXPANSION SPACE The space associated with the expansion process, and, in Stirling and similar equipment, with the absorption of heat by the working fluid undergoing expansion.

EXTERNAL COMBUSTION ENGINE An engine where the fuel is burned outside its working space, separate from the working fluid.

FLUID COMPRESSION Compression by means other than decreasing volume or increasing temperature.

FLUID FRICTION LOSS The inefficiency caused by the creation of entropy as a working fluid passes through heat transfer elements and connections.

FLYWHEEL An element of finite mass (usually a disk shape), which stores mechanical energy in its angular momentum, speeding up as it accepts energy and slowing down as it delivers energy.

FLYWHEEL PISTON A piston acting like a flywheel in that it accepts and stores mechanical energy, converting it to linear momentum, and delivers mechanical energy converting it back from momentum.

FREE PISTON (OR DISPLACER) A piston (or displacer) whose motion is controlled by working fluid forces acting on it as a result of fluid pressure and optionally the action of additional elements, such as mechanical or gas springs.

GAMMA CONFIGURATION A configuration for classifying Stirling machines consisting of two cylinders, one containing a displacer and the other a piston. Usually it has the expansion space formed between the displacer and one cylinder end, and the compression space composed of the spaces formed between the displacer and the other end of the cylinder, and the piston and its cylinder end.

HEAT ACTIVATED HEAT PUMP A thermodynamic device composed of an engine and refrigerator, operating so that the work produced by the engine is delivered to the refrigerator.

HEAT EXCHANGER The part of a real embodiment in which heat is transferred between the working fluid and the environment.

HEAT PUMP A device that absorbs heat at a lower temperature and emits heat at a higher temperature; usually referred to in this text as a refrigerator, so that the term heat pump denotes the combined engine-refrigerator.

HEAT SINK A mass or space that accepts heat that is rejected from the refrigerator, engine, or heat activated heat pump. In a space conditioning unit, this would be the outdoor environment in cooling and the living or working environment (residence, office, or other area) in heating.

HEAT SOURCE A mass or space that provides heat that is absorbed by the refrigerator, engine, or heat activated heat pump. For a refrigerator (or the refrigerator segment), this could be the outdoor environment in cooling and the living or working environment (residence, office, or other area) in heating. For an engine, this is usually the combustion products providing the heat input to drive the engine.

HEAT STORAGE REQUIREMENT The amount of heat that must be stored by a regenerator so that the working fluid passing through it enters a working space at the temperature associated with the working space.

HEATING COP, COP_h The ratio of the heat delivered at the high-temperature heat exchanger of a refrigerator to the net energy input driving the refrigeration cycle. The energy units in the numerator and the denominator are the same so COPs are dimensionless and without units.

HOT-END TEMPERATURE The temperature at which an engine or the engine segment of a heat-activated heat pump (engine-refrigerator) is absorbing heat; usually the highest heat pump temperature.

IDEAL CYCLE A theoretical (engine or refrigerator) cycle that has the best possible performance within the constraints of the Second Law of Thermodynamics.

INTEGRATED HEAT PUMP A heat pump that intimately couples the functions of the engine and refrigerator segments and shares components between them.

ISENTROPIC A type of thermodynamic process consisting of volume, temperature, and pressure changes at constant entropy.

ISOBARIC A type of thermodynamic process consisting of volume and temperature changes at constant pressure.

ISOCHORIC A type of thermodynamic process consisting of temperature and pressure changes at constant volume.

ISOTHERMAL A type of thermodynamic process consisting of volume and pressure changes at constant temperature.

LORENZ CYCLE A theoretical thermodynamic class of cycles consisting of four steps: adiabatic expansion, variable temperature compression (as a function of entropy), adiabatic compression, and variable temperature expansion (again, as a function of entropy). Specifying the function defines the particular variation.

MECHANICAL COMPRESSION Changing the pressure of a working fluid by altering its volume while accepting mechanical work.

MECHANICAL COMPRESSION CYCLE A category for engine, refrigerator, or heat pump cycles in which the device uses mechanical compression as its primary process.

MECHANICAL HEAT PUMP A heat pump with a prime mover or engine segment that produces mechanical work to drive the refrigerator segment.

NODAL ANALYSIS A type of computationally intensive analysis involving the division of the working space into a number of volumes, or nodes, to calculate the heat and mass transfer that occurs with respect to time through the operating cycle.

OPEN CYCLE A thermodynamic cycle in which at least some of the refrigerant is discharged and is not recirculated.

OPEN-SHAFT VAPOR-COMPRESSION REFRIGERATOR A real refrigerating machine operating on a Clausius-Rankine refrigerating cycle and using a compressor driven through a shaft that is exposed to the environment.

OPERATING SPEED The rate at which the thermodynamic cycle is operated.

PERFECT EQUIPMENT Hypothetical equipment whose components all operate as well as possible within the laws of thermodynamics.

PISTON A moving element that has a designed, intended pressure difference between opposite sides or faces. A piston develops or absorbs work in operation. A piston can have a temperature difference across it.

PRESSURE RATIO The ratio of the maximum pressure achieved during a cycle's operation to the minimum pressure during the same cycle.

PRIME MOVER A device that provides mechanical power, such as an engine or electric motor.

RANKINE CYCLE A theoretical cycle consisting of four steps as a refrigerator: isentropic compression, working fluid condensation, isenthalpic expansion, and working fluid evaporation. The order of the steps is reversed for an engine, with isenthalpic expansion replaced by adiabatic compression.

REAL CYCLE A cycle that describes the states of a working fluid as it passes through actual equipment.

RECUPERATIVE HEAT EXCHANGE The heat transfer process occurring in a recuperator, at or near constant pressure, between working fluids, usually in a continuous process; usually for heat recovery.

REFERENCE CYCLE A theoretical cycle that is used as a basis for comparing other cycles, particularly when calculating refrigeration efficiency.

REFRIGERANT The working fluid in a refrigerator.

REFRIGERATION EFFICIENCY, η The ratio of two COPs. The denominator is the COP of the reference cycle. The types of COPs in the numerator and the denominator should be identified by subscripts.

REFRIGERATOR A device that uses a working fluid to transfer heat from a lower temperature to a higher temperature. It receives power from an engine or other prime mover.

REGENERATIVE HEAT EXCHANGE The cyclic heat transfer process occurring in a regenerator.

REGENERATOR An element that serves as a heat reservoir. It absorbs heat as hot gas passes through it to the cold side, reducing the gas to (or near) the cold temperature; and it rejects heat as cold gas passes through it to the hot side, increasing the gas temperature to (or near) the hot temperature.

SEGMENT Either the engine or refrigerator portion of an integrated heat pump.

SINK TEMPERATURE The temperature at which the rejected heat from a refrigerator, engine, or heat pump is accepted.

SOURCE TEMPERATURE The temperature at which the heat absorbed by a refrigerator, engine, or heat pump is supplied.

SPECIFIC MASS FLOW RATE A measure of equipment size: The ratio of the mass flow rate (through components such as heat exchangers) to its working volume.

SPECIFIC POWER A measure of equipment size: The ratio of the power output of an engine to its working volume.

SPECIFIC VOLUMETRIC COOLING CAPACITY A measure of equipment size: The ratio of the heat absorbed from the low-temperature source to its working volume.

SPECIFIC VOLUMETRIC FLOW RATE A measure of equipment size: The ratio of the volumetric flow rate (through components such as heat exchangers) to its working volume.

SPECIFIC VOLUMETRIC HEATING CAPACITY A measure of equipment size: The ratio of the heat delivered from the high-temperature sink to its working volume.

SPECIFIC VOLUMETRIC REFRIGERATION CAPACITY A measure of equipment size: The ratio of the heat absorbed from the low-temperature source to its working volume; the same as specific volumetric cooling capacity, except that it is usually for a low-temperature range.

SPLIT-CYCLE MACHINE A term describing a heat pump composed of engine and refrigerator segments operating on different cycles, e.g., Stirling-Rankine.

STIRLING CYCLE An ideal thermodynamic cycle consisting of four steps as a refrigerator: isochoric expansion, isothermal expansion, isochoric compression, and isothermal compression. The order of the steps is reversed for an engine.

SWEPT VOLUME The volume that the piston or displacer traverses as it moves through (sweeps through) the cylinder it is in.

THEORETICAL CYCLE A cycle that describes the states of a working fluid as it passes through perfect equipment.

THERMAL COMPRESSION Changing the pressure of the working fluid by altering its bulk average temperature of the working fluid (usually through heat exchange), without changing the overall volume devoted to the fluid.

THERMAL COMPRESSION CYCLE A category for heat pump cycles in which the refrigerant compressor is driven directly by heat, rather than mechanical work from a separate engine segment.

THERMOCOMPRESSOR A compressor in which alternating pressure increase and decrease is achieved by the addition and rejection of heat in a constant volume working space.

TURBOCOMPRESSOR A compressor in which alternating pressure increase is achieved by imparting momentum to flowing gas by rotating blades.

TURBOEXPANDER An expansion machine in which power is produced by imparting momentum to rotating blades by flowing gas.

VAPOR COMPRESSION A real vapor compression process as part of the real Clausius-Rankine cycle.

VOID VOLUME See Dead-Space Volume.

VUILLEUMIER CYCLE A real cycle in which working fluid can be at any of three temperatures. It superficially resembles a Stirling engine and Stirling refrigerator combination with shared working fluid and instantaneous uniform pressure.

WORK STORAGE REQUIREMENT The amount of work that must be stored so that the portion of the cycle producing work can provide what is needed in the portion of the cycle absorbing work; usually a spring or a flywheel is used.

WORKING FLUID The medium that goes through the temperature and pressure variations that embody the cycle of an engine or refrigerator. Common working fluids used in integrated heat pumps are hydrogen, helium, and air.

WORKING VOLUME The volume of the heat pump that is occupied by working fluid.

WORKING VOLUME RATIO The ratio between the expansion and compression space volumes of an embodiment.

Index

Air conditioning, 4, 5, 27, 30, 41, 42
Alpha configuration, 80
Auxiliary power, 39, 84, 141

Balanced-compounded:
  Stirling heat pump, 40, 44, 58–63, 73, 92–94, 103–104, 115–118, 121, 123
  Vuilleumier heat pump, 39, 44, 63–68, 73, 97–99, 103–104, 113, 115–123
Beale, William, 43, 129, 149
Benson, Glendon M., 44–46, 68, 71, 123
Berchowitz, David M., 79
Beta configuration, 80, 95
Bounce space, 57, 58, 80
Brayton, George, 28, 42
Brayton cycle, 5, 14, 16, 28–29

Carnot, Nicolas Leonard Sadi, 20
Carnot cycle, 6, 13, 15, 19–22, 24–26, 31, 76, 77
Clausius, Rudolf J. E., 27
Clausius-Rankine cycle, 4, 14, 16, 19, 27, 33, 37, 38, 41, 43
Closed cycle, 10, 11, 27
Coefficient of performance, 14, 15, 19
Coleman Company, 2
Combined cycle, 20, 29–33
Combined engine-refrigerator cycle, 30–33
Comfort cooling, 1, 5, 6, 36
Comfort heating, 3, 6, 36, 40, 84
Compression, 9–11, 21–27, 39, 49, 54, 55, 66, 70, 76, 77, 90, 116
  mechanical, 39, 49, 53, 58, 63, 68, 84, 92–101, 116, 121, 132–134
  ratio, 95, 96
  space, 63, 65, 66, 73, 76, 77, 95, 114, 143

Compression (*Cont.*):
  thermal, 39, 42, 49, 52, 81, 84, 90–92, 120–121, 132–134, 141, 144–146
Compressors, 5, 10, 11, 14, 15, 41
Concept, 9–11
Constrained:
  displacer, 48, 57
  piston, 42, 57, 63–65, 69
Continuous flow, 22, 24, 25, 113
Cooke-Yarborough, E. H., 128
COP, 14–16
  cooling, 14–16, 21, 27, 30, 112, 113
  heating, 14–16, 21, 27–29, 112, 118
  ideal, 14–16, 31
  real, 15, 16
  theoretical, 15, 16, 27–29
Crankshaft, 44, 49, 57, 64, 65, 71, 117
Cryogenics, 6, 41–44, 125, 126
Cycle, 9
  Brayton, 5, 14, 16, 28–29
  Carnot, 6, 7, 13, 15, 19–22, 24–26, 31, 76, 77
  Clausius-Rankine, 4, 14, 16, 19, 27, 33, 37, 38, 41, 43
  Ericsson, 13, 19, 22–25, 31, 33, 44, 68, 69
  Lorenz, 4, 5, 7, 19, 26–29, 34
  Rankine, 36, 44
  Schmidt, 76–79, 82
  Stirling, 13, 19, 22, 24, 25, 31, 33, 73, 75, 76, 92
  Vuilleumier, 32–34, 39, 40, 125, 126

Danish Invention Center, 42
Dead space, 45, 52, 76, 88, 101–104, 113, 120, 121, 137, 140, 144, 147
  volume, 77, 88, 102–104, 113, 120, 138, 139, 141, 143, 144
  volume ratio, 88, 143–147

249

Design optimization, 125, 136–147
Discontinuous flow, 22, 32
Duplex Stirling heat pump, 40, 41, 43, 52–58, 94–97, 103–104, 115–117, 121, 123, 129–132
DuPré, F. K., 45, 49–74

Eder, Franz X., 81, 86, 126–128
Energy Research and Generation, Inc. (ERG), 44, 45
Engine, 9–14
  alpha-type, 80
  beta-type, 80, 95
  cycle, 10–11, 13, 21
  cylinder, 49, 52, 59, 126
  efficiency, 3, 83
  gamma-type, 80, 81, 95
Engine-driven heat pumps, 3, 11, 35, 43, 44
Equilibrium, 81, 98
Ericsson, John, 22, 44
Ericsson cycle, 13, 19, 22–25, 31, 33, 44, 68, 69
Ericsson-Ericsson heat pump, 36, 37, 44–45, 68–71, 73, 90, 99–101, 103–104, 115–118, 121, 123
Expander, 5, 10, 11, 14
Expansion, 10, 20–27, 48, 55–57, 62, 67, 71, 76, 77, 116
  space, 63, 66, 67, 68, 73, 76, 77, 80, 95, 114, 143
External combustion engine, 3, 39, 136
External drive, 73, 84

Fairchild Industries, 80
Finkelstein, Theodor, 44, 45, 63, 74, 77, 79, 85
First-order analysis, 76–79, 82, 83, 88
Fluid friction losses, 51, 54, 77, 111, 139, 141, 142, 144, 147
Flywheel, 11, 39, 63, 117
  pistons, 69, 71, 99, 100, 101, 115–117
Free piston, 123
Free-piston Vuilleumier, 63–65, 68

Gamma configuration, 80, 81, 95
Gas springs, 53, 57, 58, 71, 84
Gedeon, David R., 79–81
Gorrie, John, 28, 34

Heat-activated heat pump, 36–39, 41
Heat exchanger design, 3, 26, 136, 140, 141

Heat pump, 3, 9–14, 19, 35, 82–84
  balanced-compounded:
    Stirling, 40, 44, 58–63, 73, 92–94, 103–104, 115–118, 121, 123
    Vuilleumier, 39, 44, 63–68, 73, 97–99, 103–104, 113, 115–123
  cycle, 19, 31, 32, 34, 39
  duplex Stirling, 40, 41, 43, 52–58, 94–97, 103–104, 115–117, 121, 123, 129–132
  engine-driven, 3, 11, 35, 43, 44
  Ericsson-Ericsson, 36, 37, 44–45, 68–71, 73, 90, 99–101, 103–104, 115–118, 121, 123
  heat-activated, 36–39, 41
  integrated, 1, 4, 9, 35, 38, 39, 40, 41, 125
  mechanical, 9, 10, 36, 37, 38
  multiple-cycle, 39, 44, 45, 68, 87, 89, 121, 123
  Stirling-Stirling, 33, 43–44, 52–63, 92–97, 115, 117, 129–132
  Vuilleumier, 4, 92, 127–128, 132, 134–136
Heat sink, 15–17, 21, 24, 26, 138
Heat source, 15–17, 21, 24, 26, 130, 138
Heat storage requirement, 111–113, 121

Ideal cycle, 13–16, 20–22, 30–31, 33, 90
Integrated heat pump, 1, 4, 9, 35, 38, 39, 40, 41, 125

Joule, James, 28, 42

Kelvin, Lord, 2
  (*See also* Thomson, Sir William)
Kelvin refrigerator, 22
Khan, M. I., 77
Kinast, John A., 81
Kuehl, H. D., 81

Lorenz, H., 26
Lorenz cycle, 4, 5, 7, 19, 26–29, 34

Martini, William, 41, 85
Mass flow rate, 14
Mechanical compression, 39, 49, 53, 58, 63, 68, 84, 92–101, 116, 121, 132–134
Mechanical energy storage, 9, 57, 71, 73, 116–119
Mechanical heat pump, 9, 10, 36, 37, 38
Mechanical refrigeration cycle, 9, 15

Model validation, 80, 147
Multiple-cycle heat pump, 39, 44, 45, 68, 87, 89, 121, 123

Nodal analysis, 78–82, 85, 88

Open cycle, 10, 11, 22, 28
Open-shaft vapor-compression refrigerators, 3
Osaka Gas Company, 134

Perkins, Jacob, 27
Phasor, 71
Philips Laboratories, 42
Philips Research Laboratories, 42, 58, 125–126
Pritcher, G. K., 45, 49, 74

Qvale, E. B., 77, 80

Rallis, C. J., 79
Rankine, W. J. M., 27
Rankine cycle, 36, 44
Rauch, J. S., 80
Real cycle, 14, 15, 16
Reference cycle, 4, 13–16, 19–29
Refrigerant, 9–11, 14–15, 26–28
Refrigeration:
  cycle, 9–10, 13, 21
  efficiency, 16, 17, 26, 29
Refrigerator, 9, 10–13, 15
  cylinder, 49, 52, 61, 126
Rhombic drive, 81

Sanyo Electric Company, 134, 149
Schmidt, Gustav, 76, 85
Schmidt cycle, 76–79, 82
Schock, Alfred., 79
Schulz, S., 86, 132, 148
Second-order analysis, 76, 78, 82, 85, 87, 89–91, 118, 136, 137
Siemens, Sir William, 58
Siemens configuration, 58, 59
Smith, J. L. Jr., 77, 80
Specific mass:
  cooling capacity, 14
  flow rate, 14
  heating capacity, 14
Specific power, 83, 84
Specific volumetric:
  cooling capacity, 14, 17, 73, 115, 143, 144
  flow rate, 14

Specific volumetric (*Cont.*):
  heating capacity, 14, 17, 73, 115
  refrigeration capacity, 39
Split-cycle machines, 37, 44
Staged refrigerators, 30
Stirling, Robert, 22, 75
Stirling:
  balanced-compounded, 40, 44, 58–63, 73, 92–94, 103–104, 115–118, 121, 123
  cycle, 13, 19, 22, 24, 25, 31, 33, 73, 75, 76, 92
Stirling-Stirling heat pump, 33, 43–44, 52–63, 92–97, 115, 117, 129–132
Sunpower, Inc., 43, 79, 80, 85, 129–131
Swash-plate drive, 81
Swept volume, 132, 143, 147

Technical University of Munich, 126–127, 147
Theoretical cycle, 13–16, 25–29
Thermal compression, 39, 42, 49, 52, 81, 84, 90–92, 120–121, 132–134, 141, 144–146
  Vuilleumier, 91–92, 94, 98, 103–104, 111–112, 115, 116, 120, 121, 141, 144
Thermally driven compressor, 11
Thermizers, 45
Thermocompressor, 41, 43
Thermoelectric devices, 9, 36
Third-order analysis, 76, 78, 88
Thomson, Sir William, 2, 7, 34
  (*See also* Kelvin, Lord)
Toho Gas Company, 134
Tokyo Gas Company, 134
Turbocompressor, 10, 20, 21, 24
Turboexpander, 20, 21, 24
Turbomachinery, 5, 22

University of Dortmund, 132
Urieli, Israel, 79, 81

Van Weenan, F. L., 58
Vapor compression, 2–5, 27–28, 43
Vuilleumier, Rudolph, 4, 31, 41
Vuilleumier:
  balanced-compounded, 39, 44, 63–68, 73, 97–99, 103–104, 112, 113, 115–123
  cycle, 32–34, 39, 40, 76, 125, 126
  free-piston, 63–65, 68
  heat pump, 4, 92, 127–128, 132, 134–136

Vuilleumier (*Cont.*):
   with internal heat exchangers, 39, 43, 51–52, 73, 90–92, 103
   thermal compression, 91–92, 94, 98, 103–104, 111–112, 115, 116, 120, 121, 141, 144
   traditional, 39, 41–43, 48–51, 73, 90–92

Walker, Graham, 43, 45, 46, 77, 85
Work storage, 116, 118, 123
Working fluid, 3–5, 9, 10, 13, 14, 16, 17, 21, 36–40, 42, 43, 104–111, 136
Working volume ratio, 143, 144, 147

Zero-order analysis, 76, 77, 82

About the Authors

JAROSLAV WURM is assistant director of space conditioning research at the Institute of Gas Technology with responsibility for the development and evaluation of new refrigeration and space conditioning technologies and other energy transfer systems.

JOHN A. KINAST is engineering supervisor of space conditioning research at the Institute of Gas Technology where he has applied computer simulation methods to the study of heat pumps and is working on the development of Stirling engine-driven gas-fired heat pumps.

DR. THOMAS R. ROOSE is a principal scientist at the Gas Research Institute with management responsibilities including heat pumps, heat and mass transport, and combustion.

DR. WILLIAM R. STAATS is director of the physical sciences department at the Gas Research Institute. He is in charge of the national program of basic research for the United States natural gas industry.